Ecological Genetics
and Evolution

Proceedings of a Workshop held at the
University of Arizona Conference Center, Oracle, Arizona on January 4–8, 1982
under the auspices of the U.S./Australia Cooperative Science Program

Ecological Genetics and Evolution

The Cactus-Yeast-*Drosophila* Model System

Edited by

J. S. F. Barker

Department of Animal Science
University of New England

W. T. Starmer

Department of Biology
Syracuse University

ACADEMIC PRESS

A Subsidiary of Harcourt Brace Jovanovich, Publishers

Sydney   New York   London
Paris   San Diego   San Francisco   São Paulo   Tokyo   Toronto
1982

ACADEMIC PRESS AUSTRALIA
Centrecourt, 25–27 Paul Street North
North Ryde, N.S.W. 2113

United States Edition published by
ACADEMIC PRESS INC.
111 Fifth Avenue
New York, New York 10003

United Kingdom Edition published by
ACADEMIC PRESS, INC. (LONDON) LTD.
24/28 Oval Road, London NW1 7DX

Printed in Australia

National Library of Australia Cataloguing-in-Publication Data

Ecological genetics and evolution.

  Includes bibliographies and index.
  ISBN 0 12 078820 9.

  1. Evolution - Congresses. 2. Ecological
genetics - Congresses. I. Barker, J.S.F.
(James Stuart Flinton), 1931-. II. Starmer, W.T.
III. Title : The cactus-yeast-Drosophila model system.

575.1'3

Library of Congress Catalog Card Number: 82-72224

Academic Press Rapid Manuscript Reproduction

Contents

PART IV   POPULATION GENETICS AND ECOLOGY

Contributors

Numbers in parentheses indicate the pages on which the authors' contributions begin.

J. S. F. Barker (209), Department of Animal Science, University of
New England, Armidale, New South Wales 2351

Philip Batterham (307), Department of Biology, Syracuse University, Syracuse,
New York 13210

P. D. East (323), Department of Animal Science, University of New England,
Armidale, New South Wales 2351

James C. Fogleman (191), Department of Ecology and Evolutionary Biology
University of Arizona, Tucson, Arizona 85721

Antonio Fontdevila (81), Departamento de Genética, Facultad de Ciencias,
Universidad Autónoma, Bellaterra (Barcelona), Spain

Arthur C. Gibson (3), Department of Biology, University of California,
Los Angeles, Los Angeles, California 90024

John B. Gibson (291), Department of Population Biology, Research School
of Biological Sciences, The Australian National University,
Canberra, A.C.T. 2601

William B. Heed (65), Department of Ecology and Evolutionary Biology,
University of Arizona, Tucson, Arizona 85721

D. L. Holzschu[1] (127), Department of Food Science and Technology,
University of California, Davis, California 95616

J. Spencer Johnston (225, 241), Department of Plant Sciences,
Texas A&M University, College Station, Texas 77840

Henry W. Kircher (143), Department of Nutrition and Food Science,
University of Arizona, Tucson, Arizona 85721

R. L. Mangan[2] (257), Department of Entomology, The Pennsylvania State
University, and U.S. Regional Pasture Research Laboratory, University Park,
Pennsylvania 16802

Therese Ann Markow (273), Department of Zoology, Arizona State University,
Tempe, Arizona 85287

N. D. Murray (17), Department of Genetics and Human Variation,
La Trobe University, Bundoora, Victoria 3083

John G. Oakeshott (291), Department of Population Biology, Research School
of Biological Sciences, The Australian National University,
Canberra, A.C.T. 2601

Present addresses

[1]Department of Bacteriology, University of California, Davis, California 95616

[2]USDA Screwworm Research, Mission, Texas

M. A. Q. R. Pereira (97), Instituto de Biociências, Universidade de São Paulo, São Paulo, Brasil CEP 05421

H. J. Phaff (127), Department of Food Science and Technology, University of California, Davis, California 95616

R. H. Richardson (107), Department of Zoology, University of Texas at Austin, Austin, Texas 78712

F. de M. Sene (97), Instituto de Biociências, Universidade de São Paulo, São Paulo, Brasil CEP 05421

William T. Starmer (159, 307), Department of Biology, Syracuse University, Syracuse, New York 13210

David T. Sullivan (307), Department of Biology, Syracuse University, Syracuse, New York 13210

Alan R. Templeton (225, 241), Department of Biology, Washington University, St Louis, Missouri 63130

Lynn H. Throckmorton (33), Department of Biology, University of Chicago, Chicago, Illinois 60637

Don C. Vacek (175), Department of Animal Science, University of New England, Armidale, New South Wales 2351

C. R. Vilela (97), Instituto de Biociências, Universidade de São Paulo, São Paulo, Brasil CEP 05421

Marvin Wasserman (49), Biology Department, Queens College, Flushing, New York 11367

Preface

Ecological genetics and evolutionary biology are subjects which can be fully developed only through a multidisciplinary approach. Thus theoretical and empirical studies, ranging from biogeography and systematics through ecology to protein chemistry and enzyme kinetic analyses, are necessary components. But integration of these studies is essential if a comprehensive understanding of the genetic strategies of adaptation is to be gained. While study of the interactions between the genetic structure of populations and the temporal and spatial variations presented by the environment has provided much of our current understanding, integration can only be achieved through concerted study of particular biological systems. Naturally, the knowledge gained from studying any particular system should be capable of extension to ecological genetics and evolution in general.

Within the last ten years, extensive research programs have been initiated in Australia, Brasil, Spain and the USA, all of which focus on the same biological system for studying aspects of population genetics, ecology, systematics and evolution. This system involves species of *Drosophila* and yeast which live in the decaying stems, fruits and cladodes of various cacti. The decaying cactus tissue is a nutrient-rich environment for the growth of of micro-orgnaisms, which in turn supply essential nutrients for the *Drosophila* . The micro-organisms rely on insects for dispersal to new habitats, and the *Drosophila* serve (at least partially) this vectoring role.

The elements which make this a model system for studies in population biology and evolution are:

(1) The complementary studies of the biogeography and systematics of the three components, *viz.* cacti, micro-organisms and *Drosophila*, which permit clearer interpretation of speciation in, and the evolution of, each component.

(2) The specific knowledge of the host plants that support various species of yeasts and *Drosophila*, which allows field investigations to be conducted at the natural breeding sites and provides information on similarities and differences among habitats. Comparisons of habitat characteristics (including the chemistry of the cacti) can and have led to testable hypotheses concerning the evolution and adaptation of the host specific

organisms. Natural selection affecting gene frequencies at enzyme loci in natural populations has been demonstrated for the first time in this system.

(3) The ability to grow and study the yeasts and *Drosophila* in the laboratory. This aspect of the model system provides information on the genetic composition of the *Drosophila* populations (both chromosomal and electrophoretic variation) and the physiological abilities of the yeasts. Moreover, laboratory experiments for testing theoretical questions are possible.

(4) The ability to translate the results from laboratory investigations to the natural field conditions.

The state of study of cactus specific organisms has reached a rich level and is now ripe for extension into new areas. It therefore seemed timely for workers utilizing this system to meet and review the current parallel and complementary research programs.

In addition to the papers presented, five Workshop Discussion Sessions were devoted to specific topics in order to review existing knowlege and research in progress, to identify perceived deficiencies in our knowledge and to plan future research. Summaries of the discussion and conclusions from these sessions are included.

We wish to record our thanks to the Department of Science and Technology, Australia, and to the National Science Foundation, USA, for the provision of financial support under the U.S./Australia Cooperative Science Program, and to Instituto Nacional de Asistencia y Promocion del Estudiante, Ministerio de Educacion y Ciencia, Spain, and to FAPESP and CNPq, Brasil, for financial support for Dr A. Fontdevila and Dr F. de M. Sene, respectively. Our special thanks go to Professor W. B. Heed, University of Arizona, for local arrangements, to the staff of the University of Arizona Conference Center for their assistance during the meeting, to Peter East, Don Gilbert and Don Vacek for assistance in preparing these Proceedings for publication, and to Dorothy Cordingley and Jill Parker for typing, cheerfulness and patience in assisting with the preparation of the final manuscripts and index.

PART I
THE HOST CACTI

1
Phylogenetic Relationships of Pachycereeae

Arthur C. Gibson

Department of Biology
University of California, Los Angeles
Los Angeles, California

I Introduction

In North America, *Drosophila* of the species groups
repleta and *nannoptera* lay eggs and feed in the rotting
succulent stems of columnar cacti belonging to tribe
Pachycereeae; and drosophilids inhabiting these columnar
cacti tend to be specific to one or several host species
(Fellows and Heed, 1972). Because the columnar cactus-
microorganism–*Drosophila* relationship is a superb system in
which to study ecological genetics, a desirable goal is to
compare the phylogenies of the flies with one constructed for
their host plants. Viewing this optimistically, such comparisons may reveal not only the origins of deserticolous
forms but also some structural and physiological bases for
their host specificity.

Even though the systematics of Cactaceae is still very
unsettled and confusing, there is a fairly broad consensus on
what groups should be assigned to tribe Pachycereeae. All
species have a wood skeleton consisting of a ring of discrete, parallel rods, which may eventually fuse laterally in
old stems (Gibson, 1978). Tribe Pachycereeae includes approximately 70 species, of which 90% are endemic or nearly so to
Mexico. Species range northward into central Arizona, southward to coastal and northern South America, and eastward into
the West Indies and southernmost Florida. Simple comparisons
suggest that Pachycereeae are most closely related to
Leptocereeae of northwestern South America, especially
Armatocereus.

ECOLOGICAL GENETICS AND EVOLUTION
ISBN 0 12 078820 9

A survey was conducted on Mexican columnar cacti as well
as outgroups of columnar forms to describe and interpret the
diverse anatomy of Pachycereeae and to develop a phylogenetic
classification of the tribe (Gibson and Horak, 1978).   In
this paper, the pertinent systematic and biogeographic observations on Pachycereeae are summarized to aid research on the
ecology of *Drosophila* and cactophilic yeasts.   Readers should
consult the original article for all detailed descriptions
and lists of materials used.

II   Observations on Stem Anatomy

Typical stem structure of a columnar cactus is illustrated in Figure 1.   Chlorophyll-bearing succulent tissue
(chlorenchyma) of the cortex is covered by a tough but flexible skin, which consists of an epidermis with a waxy cuticle
and a thick-walled, collenchymatous hypodermis.   Collenchyma
has much pectin and hemicellulose in its primary walls, so
the tissue is markedly hygroscopic.   Hypodermis is inter-
rupted at various points by canals, extending from the
stomates to the chlorenchyma to permit gas exchange.
An exciting discovery was the finding of silica bodies
(opals) in the hypodermis and epidermis of numerous
Pachycereeae (Fig. 2).   Silica bodies are very uncommon in
angiosperms as a whole and have not been observed in other
Cactaceae, making this an unusual apomorphic feature.   The
majority of species once classified by Britton and Rose
(1920) in *Lemaireocereus* plus their genus *Machaerocereus* and
Rathbunia kerberi definitely have silica bodies; but they are
absent from other Mexican species once assigned to
Lemaireocereus, such as *hollianus*, *marginatus* and *weberi*.
Species of *Cephalocereus* from North America as well as
Mitrocereus fulviceps and several species of *Neobuxbaumia*
have many small rhomboidal crystals of calcium oxalate in
each epidermal cell and large solitary ones in the hypodermis.   This crystal pattern apparently occurs nowhere else
in Cactaceae, making this an apomorphic feature.   Silica
bodies and calcium oxalate crystals in the skin are mutually
exclusive.
Figure 1 also shows the presence in the cortex of mucilage cells, which are common in most highly specialized
cacti.   Quite unexpectedly, this survey revealed that mucilage cells are absent from the succulent cortex and pith in
Polaskia chende (syn. *Heliabravoa* or *Lemaireocereus chende*),
Polaskia chichipe, *Pterocereus gaumeri*, *Pachycereus*

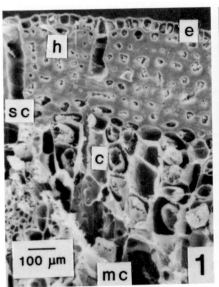
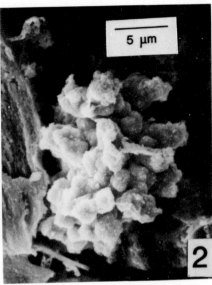

FIGURE 1. *Transection of the outer stem of agria, Stenocereus gummosus, as seen with the scanning electron microscope. This section shows the epidermis (e), a thick hypodermis (h) with conspicuous substomatal canals (sc), and chlorenchyma (c) with scattered mucilage cells (mc).*

FIGURE 2. *A high magnification view of a large coralloid silica body in the hypodermis of Stenocereus gummosus. Only one silica body occurs in each cell.*

marginatus, and *Pachycereus* or *Lemaireocereus hollianus*. This is significant because species of tribe Leptocereeae, the putative plesiomorphic outgroup, also lack stem mucilage cells in certain species, reinforcing a subjective judgment that this should be the ancestral condition. Moreover, in *Polaskia* and *Pterocereus* the hypodermis is thin and has scanty collenchymatous walls, as expected in fairly unspecialized forms. In contrast, those taxa like organpipe (*Stenocereus thurberi*), which has cortical mucilage sacs that are clearly visible without a hand lens, and species with many, closely distributed mucilage cells, such as cina (*Stenocereus alamosensis*), are considered highly derived on many morphological grounds. Several species have small, scattered mucilage cells in the cortex but none in the pith, whereas highly derived columnar cacti usually have mucilage cells in the pith. Assuming that mucilage has an important function in increasing matrix forces for retaining water and possibly in acting as a deterrent to herbivores, one can hypothesize a nonreversible trend from none to many mucilage structures. In this vein, most of the primitive-looking leaf-bearing species of *Pereskia* also have few, if any, mucilage cells in the shoot (Bailey, 1964).

Whereas some character states are unusual enough to allow us to make defensible statements on the direction of evolution for character states, numerous other features are more ambiguous. Species of *Stenocereus* with very similar types of cuticle also share some more subtle structural characteristics of the epidermis and hypodermis. In another group of species (*Backebergia* and *Pachycereus*), the outer cortical cells are greatly elongate in the radial direction. Nonetheless, character states like multi-layered epidermis, thick hypodermis, and sunken stomates appear not only in highly derived species of *Pachycereus* but also in very dissimilar species, such as organpipe (*Stenocereus thurberi*). Thus, other characters than those of stem anatomy must be sought to estimate the phylogenetic relationships of these species.

III   Chemistry of the Stem

Stem chemistry of cacti has long been of interest to phytochemists because certain species are known to have hallucinogenic compounds. For this reason, the chemistry of Pachycereeae has been studied in some detail. Table I lists the species known to have alkaloids (subtypes tetrahydro-isoquinolines and tyramine-types) and glycosidic triterpenes (acidic oleanane and lupane types). Alkaloids and glycosidic triterpenes are mutually exclusive in any species. Although one or two compounds may be present in high concentrations, numerous species have several additional compounds in significant quantities; and future work will undoubtedly find numerous minor compounds as well. Although Gibson and Horak (1978) suggested that these compounds may be important to discourage foraging and the growth of microorganisms, experiments are needed to demonstrate clearly how any of these roles would have affected the early evolution of these species. No cactus research has attempted to unravel why several compounds occur in a single stem, although this may be a way to inhibit organisms that may be able to tolerate one type of alkaloid or triterpene but not another.

TABLE I. Species of Pachycereeae Known to have Tyramine and Tetrahydroisoquinoline Alkaloids or Glycosidic Triterpenes

Tyramine and tetrahydroisoquinoline alkaloids

Backebergia militaris, Carnegiea gigantea, Lophocereus gatesii, L. schottii, Pachycereus marginatus, P. pecten-aboriginum, P. pringlei, P. weberi

Glycosidic triterpenes (oleanane series)

Escontria chiotilla, Myrtillocactus cochal, M. eichlamii, M. geometrizans, M. schenckii, Polaskia chende, P. chichipe, Stenocereus alamosensis, S. beneckei, S. dumortieri, S. griseus, S. gummosus, S. hystrix, S. longispinus, S. montanus, S. pruinosus, S. queretaroensis, S. quevedonis, S. stellatus, S. thurberi, S. treleasei

Glycosidic triterpenes (lupane series)

Myrtillocactus schenckii, Stenocereus eruca, S. hystrix, S. quevedonis, S. stellatus, S. thurberi, S. treleasei

The same species that have acidic glycosidic triterpenes also show an unusual wound response by turning bright reddish-orange before fading to the dry condition; whereas plants rich in alkaloids turn reddish and then black rather suddenly and remain black in the dry condition.

IV Features of the Ovary

Buxbaum (1962) was first to observe that fruit pulp color in numerous Pachycereeae is produced within epidermal cells on the surface of the funiculus. These appear prior to anthesis; and long after fertilization but prior to fruit ripening, these cells ("pearl cells") become highly vacuolate, enlarge, and produce the pigment of the fruit. In contrast, other species lack these funicular pigment cells, and fruit color is produced generally within the fruit wall and diffuses into the pulp late in ontogeny. Table II shows that funicular pigment cells occur in the same species with

TABLE II.   *Species of Pachycereeae Known to have
Funicular Pigment Cells at Anthesis*

Escontria chiotilla	Stenocereus gummosus
Myrtillocactus cochal	S. hystrix
M. geometrizans	S. montanus
M. schenckii	S. pruinosus
Polaskia chende	S. quevedonis
P. chichipe	S. stellatus
Stenocereus alamosensis	S. thurberi
S. dumortieri	S. treleasei
S. eruca	

stem glycosidic triterpenes, many of which are also the
species with silica bodies. They are shared derived character states, i.e., synapomorphies.

When the ovary is sliced in longitudinal view to obtain
observations on funicular pigment cells, one can observe that
betalain pigments occur in various places in the future fruit
wall. In *Polaskia, Escontria,* and *Myrtillocactus* red pigmentation is present in the thick pericarpel and absent from the
ovary wall and beneath the locule, whereas in *Stenocereus* the
reverse is true. Green coloration generally prevails in
other genera of Pachycereeae. These same sections also
demonstrate that species with few or no mucilage cells in the
stem tend to lack them in the developing fruit wall.

There are many derived characteristics appearing in the
flowers and fruits that help to elucidate the groups of
closely related species. Most of these need careful quanti-
fication, but there are some which are quite diverse, such as
external testal features of seeds. In seeds ranging from
5 mm (*Backebergia* and *Pachycereus*) to less than 0.5 mm long
(*Escontria, Myrtillocactus,* and *Polaskia*), one observes sur-
faces that are smooth to rough, shiny to dull, reddish black
to brown or pure black. Features of the hilum and the raphe
show especially interesting diversity. Species with glycosidic triterpenes and funicular pigment cells but which lack
silica bodies have the smallest seeds. For example, seeds of
Escontria and *Myrtillocactus* are nearly identical in all
features, and these are the two genera of this group with
mucilaginous stem cortex and very thick skin. In the
alkaloid-bearing species, seeds of *Pachycereus* and
Lophocereus are shiny and black and some have a prominent
raphe. Current studies using scanning electron microscopy
will help to document the cellular and cuticular similarities
and differences of the seeds of Pachycereeae.

V Systematic Conclusions

Comparative anatomy and a reevaluation of the traditional features permitted Gibson and Horak (1978) to propose a new classification system of Pachycereeae (Table III, Fig. 3), one with a phylogenetic perspective. This is only a first approximation of the relationships of the genera and species groups, and more information is needed to develop a phylogenetic hypothesis that predicts all the relationships of the species.

The clearest group to define is the reconstituted subtribe Stenocereinae. Subtribe Stenocereinae sensu Buxbaum (1958 *et seq.*, see Gibson and Horak, 1978) included all species of *Lemaireocereus* Britton and Rose in North America except *L. hollianus* and those now assigned to *Armatocereus*, *Backebergia*, *Carnegiea*, *Lophocereus*, *Machaerocereus*, *Neobuxbaumia*, and *Rathbunia*. Now the subtribe includes only those species that have either glycosidic triterpenes or funicular pigment cells or both, i.e., the genera *Escontria*, *Myrtillocactus*, *Polaskia* (here including *Heliabravoa*), and *Stenocereus* (including *Machaerocereus*, *Rathbunia*, and other segregate genera but excluding those lacking triterpenes and funicular pigment cells). The remaining genera, *Carnegiea*, *Pterocereus*, *Neobuxbaumia*, *Backebergia*, *Mitrocereus*, and *Cephalocereus*, are tentatively assigned to another subtribe, Pachycereinae.

Within subtribe Stenocereinae, two groups of genera are identifiable, those with silica bodies, which are now called *Stenocereus*, and those lacking them. *Stenocereus* has undergone extensive evolution in all features. In the stems, changes observed include modifications in the thickness of the hypodermis, size and position of the silica bodies, and development of mucilage structures. Flowers have floral tubes ranging from 2.5 to 10 cm in length, with various widths; and color varies from white, cream, pale pink, or lavender changing to darker shades to purple red, and in the rathbunias the flowers are terracotta red. Other variable floral features include the number of flowers per areole, position and orientation of flowers, time of anthesis, spination and vestiture on the ovary at anthesis, number of stamens and whether they are exserted or inserted, number of stigma lobes, size and degree of exposure of the nectary chamber, and odor of the flower. Although much work is needed to understand the pollination biology of Pachycereeae, this is the clearest example to show that speciation has been promoted by modifications toward special large pollinators,

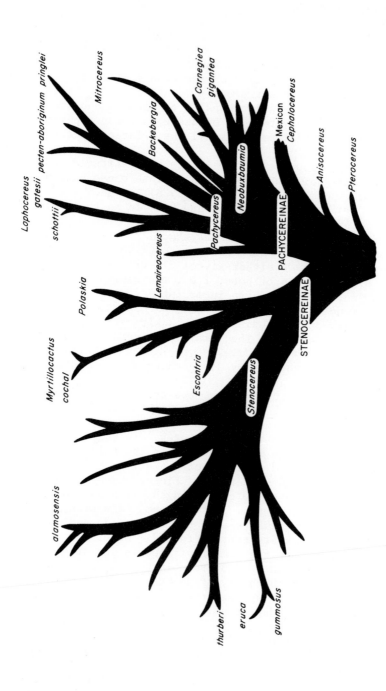

FIGURE 3.   *Tentative phylogenetic reconstruction of Pachycereeae, in which two subtribes, Stenocereinae and Pachycereinae, are recognized. Species with the majority of plesiomorphic features are placed in the center, and species native to the Sonoran Desert are named.*

TABLE III. *Proposed Classification of Tribe*
Pachycereeae. Species are not Listed for
the Largest Genera, Stenocereus and
Cephalocereus

Tribe Pachycereeae

Subtribe Stenocereinae
Polaskia (chende, chichipe)
Escontria (chiotilla)
Myrtillocactus (cochal, eichlamii, geometrizans,
schenckii)
Stenocereus (incl. Hertrichocereus, Isolatocereus,
Machaerocereus, Marshallocereus, Rathbunia, and
Ritterocereus)

Subtribe Pachycereinae
Pterocereus (foetidus, gaumeri)
Lemaireocereus (hollianus)
Pachycereus (grandis, marginatus, pecten-aboriginum,
pringlei, weberi)
Lophocereus (gatesii, schottii)
Backebergia (militaris)
Mitrocereus (fulviceps)
Neobuxbaumia (euphorbioides, macrocephala, mezcalaensis,
multiareolata, polylopha, scoparia, tetetzo)
Carnegiea (gigantea)
Cephalocereus (incl. Haseltonia and Neodawsonia but
excl. Subpilosocereus)

---

such as bats, hummingbirds, hawkmoths, and so forth. The
segregate genera are by and large extreme forms for highly
specialized pollinators, such as hummingbirds in the
Rathbunia-type and probably hawkmoths in the former
Machaerocereus gummosus.

The small-seeded species of Stenocereinae form two
clades, the genus *Polaskia*, which has more primitive character states than any other genus in the tribe, and the more
derived *Escontria* and *Myrtillocactus*, all of which have
relatively small flowers. Authors may wish to argue whether
this group of seven species is its own subtribe, but only
after full data are collected and analyzed can such decisions
be made with confidence. Certainly, the two species of
Polaskia have all the proper character states to be intermediate between the two subtribes and are ancestral (less
modified flowers) to either *Stenocereus* or

Escontria-Myrtillocactus lineages. A reason for stating this is that flowers of *Polaskia chende* have spines, bristles, and trichomes; so this could have been modified to the spiny fruits of *Stenocereus* as well as the bristly fruits of *Pachycereus* and *Backebergia* or the more naked fruits (as in *Polaskia chichipe*). *Polaskia* also has biochemically very simple triterpenes, whereas those in specialized species may be highly derived (Fig. 4).

The use of stem chemicals and crystals as well as structure of the stems, flowers, fruits, and seeds produces an entirely new classification of the remaining genera. The three characteristic species of *Pachycereus* with strong bristles on the fruits are combined with three species formerly assigned by authors to *Lemaireocereus, marginatus, weberi,* and *hollianus* (all had been classified in *Pachycereus* from time to time). In this group, the least specialized form is *Pachycereus hollianus*, and future research may show that this is distinctive enough to be classified as its own genus. If so, this would become a monotypic genus *Lemaireocereus* because this is the type of that original complex, which is now segregated into numerous genera, subtribes and even tribes. *Lophocereus* is clearly related to *Pachycereus marginatus* on the one hand and *Backebergia militaris* on the other, species with very similar alkaloids (Mata and McLaughlin, 1980), stem structure, and many reproductive features.

Saguaro (*Carnegiea gigantea*) belongs to a group of species which is mostly unknown structurally and chemically. *Carnegiea* is certainly a very close relative of *Neobuxbaumia* (Gibson and Horak, 1978; Glass and Foster, 1979), which occurs in southern Mexico, and these are in turn related to *Mitrocereus*. When full comparisons between *Carnegiea* and *Mitrocereus* are made, it is likely the two genera will have to be combined, because they are so similar in many fundamental ways. Fortunately for *Drosophila* workers, the name *Carnegiea* would be used for all species because this has priority over the younger name of the southern forms.

Difficulties arise in interpreting *Cephalocereus*, which has many features shared with *Mitrocereus* and *Neobuxbaumia*, but it is widespread and has an evolutionary history outside Mexico. The presence of calcium oxalate crystals in the skin of all species analyzed from Mexico, including the segregate *Neodawsonia*, suggests that the species observed are monophyletic, but we cannot be certain whether this clade arose from Mexican forms or is, instead, a separate lineage from northern South America. Extensive research is needed to clarify these problems.

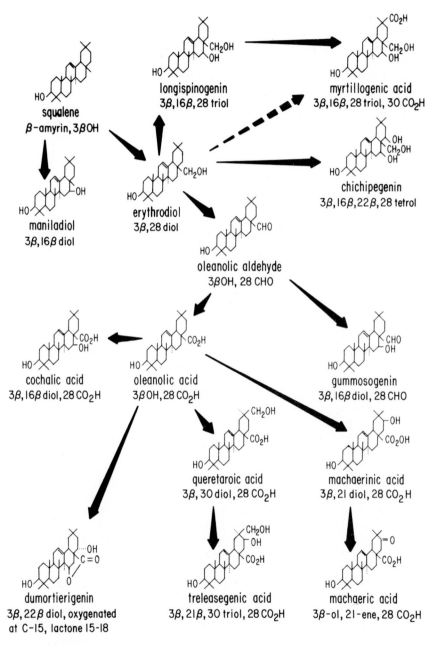

FIGURE 4. *Hypothesized biochemical phylogeny of oleanane triterpenes in Pachycereeae, derived from squalene. Substantial technical advice was made by Henry Kircher.*

Tribe Pachycereeae presently includes the unusual and fairly primitive genus *Pterocereus*. In the future, *Pterocereus* may be classified as its own monotypic subtribe if kept as a member of this tribe. Stems are relatively unspecialized, lacking mucilage cells and hypodermal crystals, but the flowers are specialized for bat pollination and have green leafy appendages. Presently, *Pterocereus* is more easily interpreted as an early, specialized offshoot of the tribe rather than the ancestor of either or both large subtribes.

VI   Origins of Columnar Cacti of the Sonoran Desert

Even when these simple preliminary changes are made, any systematist can realize that the cacti of the Sonoran Desert have come from numerous separate invasions of southern clades. The minimum number of invasions include the following: *Myrtillocactus cochal* from *M. geometrizans*; *Stenocereus thurberi* from the species of *Stenocereus* with red areolar trichomes in Sinaloa and southward; *Stenocereus gummosus* and *S. eruca* perhaps from *S. fricii* in Nueva Galicia of central western Mexico; *Lophocereus* from *Pachycereus marginatus* or *Backebergia militaris* in areas with *S. fricii*; *Carnegiea gigantea* from *Neobuxbaumia* of Puebla and southward; *Stenocereus alamosensis* from *S. kerberi* and *S. standleyi* of western Mexico; and *Pachycereus pringlei* and *P. pectenaboriginum* from western Mexico. This allows workers to predict where the related *Drosophila* and yeasts of the Sonoran cactus ecosystem should be found. Moreover, certain patterns are strikingly reminiscent of accounts on the origin of Baja California as an event of plate tectonics within the last four million years (Gastil *et al.*, 1981), and others can be attributed to the former widespread distributions of the saguaro, organpipe, rathbunia, cardon, and hecho lineages from southern Mexico along the western coastline into the present-day location of the Sonoran Desert. *Myrtillocactus* also could have had a former distribution between the northern Chihuahuan Desert into Sonora and Baja California, rather than a distribution along the western side of the Sierra Madre Occidental. In fact, distributions of the species of *Drosophila* for each of these lineages may help to unravel the probable routes of dispersal for the cacti. Long-distance dispersal is possible for these cacti, but such an explanation is not parsimonious if the flies and yeasts using the cactus substrate have the identical disjunction pattern (Heed, 1982). Instead, a more reasonable approach would be to assume that these cactus distributions were essentially continuous in the past.

The dispersal routes of Pachycereeae that could have led to the differentiation of present-day species can be hypothesized. Major clades all occur in Puebla and Oaxaca, especially in the Valley of Tehuacán, and one could assume that this region was the center of radiation because the species with plesiomorphic features occur there and their more highly derived species radiate from there northward along western Mexico or eastern Mexico or into the Caribbean region, producing species through geographic speciation. Because speciation here apparently proceeded in response to selection by different pollinators, whose ages are mostly estimated at Tertiary in age, a post-Oligocene radiation is suggested. This time-scale is consistent also with speciation events of cactophilic *repleta*, which were probably radiating in Oligocene and Miocene times (Throckmorton, 1974, 1982). Because the platyopuntias (*Opuntia*) may be as old or older than these columnar cacti, there is no reason to rule out the possibility that evolution of cactophilic *Drosophila* could have started on either and moved into the other host system.

REFERENCES

Bailey, I.W. (1964). *J. Arnold Arbor. 45*, 374.
Britton, N.L., and Rose, J.N. (1920). "The Cactaceae", Vol. 2. Carnegie Institute Publication 248, Washington.
Buxbaum, F. (1958). *Madroño 14*, 177.
Buxbaum, F. (1962). *In* "Die Kakteen" (H. Krainz, ed.). Franckh'sche Verlagshandlung, Stuttgart.
Fellows, D.P., and Heed, W.B. (1972). *Ecology 53*, 850.
Heed, W.B. (1982). Chapter 5, this Volume.
Gastil, G., Morgan, G., and Krummenacher, D. (1981). *In* "The Geotectonic Development of California, Rubey Vol. 1" (W.G. Ernst, ed.), p. 824. Prentice-Hall, Englewood Cliffs.
Gibson, A.C. (1978). *Cactus Succ. J., Gt Br. 40*, 73.
Gibson, A.C., and Horak, K.E. (1978). *Ann. Missouri Bot. Gard. 65*, 999.
Glass, C., and Foster, R. (1979). *Cactus Succ. J., Los Ang. 51*, 25.
Mata, R., and McLaughlin, J.L. (1980). *J. Pharm. Sci. 69*, 94.

Throckmorton, L.H. (1974). *In* "Handbook of Genetics", Vol. 3 (R.C. King, ed.), p. 421. Plenum, New York.
Throckmorton, L.H. (1982). Chapter 3, this Volume.

2
Ecology and Evolution of the *Opuntia-Cactoblastis* Ecosystem in Australia

N. D. Murray[1]

Department of Genetics and Human Variation
La Trobe University
Bundoora, Victoria

I Introduction

The ecological and evolutionary responses of organisms to new environments have long been a focus of interest in the study of introduced plants and animals. In particular, the special importance of both deliberate and accidental introductions as genuine experiments in population biology has been stressed by Waddington (1965) and Myers (1978). The worldwide spread of the Cactaceae following European contact with the New World provides an extensive arena for such studies, because many cactus-specific animals and microbes have subsequently been transplanted widely in attempts to control the more weedy cacti. The attendant knowledge of dates, numbers, and localities of these introductions makes them particularly valuable as evolutionary experiments.

I will review here the status of cacti and cactus-feeding insects within Australia, and describe some of the ecological and evolutionary interactions between *Opuntia* spp., the phycitid moth *Cactoblastis cactorum* (Berg), and the Australian environment.

[1]*Supported by Australian Research Grants Committee*

ECOLOGICAL GENETICS AND EVOLUTION
ISBN 0 12 078820 9

II Cacti and Associated Insects in Australia

Most cacti presently established in Australia were introduced in the nineteenth century. The relative water-
efficiency of the weedier cacti permitted rapid invasion of
rangeland and native forest during and after severe droughts
in the late nineteenth century (Dodd, 1940; Mann, 1970;
Osmond and Monro, 1981).

Table I shows that cacti established in Australia have
geographically diverse origins. The common pest species in
south and central Queensland (Lat. 23°-27°S) include *O.
stricta* from Florida and Texas, *O. tomentosa* and *O.
streptacantha* from Mexico, and *O. aurantiaca*, *O. vulgaris*,
and *Eriocereus* (=*Harrisia*) *martinii* from South America.
Species occurring in central Victoria (Lat. 37°S) are *O.
stricta*, *O. vulgaris*, the Mexican *O. robusta*, and, more rarely,
O. cylindrica, presumed to come from the equatorial highlands
of Peru and Ecuador. The Australian localities can be
climatically different from the native ones, e.g. *O. stricta*
now lives at much higher latitudes than in the New World, and
the extent to which genetic differentiation has occurred and
is occurring requires study.

Cactus-feeding insects also represent a mixture of
southern and northern hemisphere species. Some of the smaller
forms presumably entered Australia in company with the original cactus introductions, e.g. the scale insect *Diplacaspis
echinocacti* (Mann, 1970), or accidentally at a later date,
e.g. *Drosophila buzzatii* (Barker and Mulley, 1976). Table II
lists those species that have been deliberately and successfully introduced as part of the sustained campaign of
biological control mounted against the pest-species of *Opuntia*
and *Eriocereus* since 1903. Included in the table are known
host-plants of these insects in both native and Australian
habitats.

In most cases of successful establishment there is a
striking correspondence between the original host-plant and
the type of host adopted within Australia. This can be seen
for species attacking *Eriocereus*, and for scale insects
(*Dactylopius*) which generally attack the same form of *Opuntia*
as of their original hosts. *Cactoblastis cactorum* utilizes
parallel ranges of platyopuntias on the two continents: in
Australia these are the tree-pears *O. tomentosa* and *O.
streptacantha*, a shrub-pear *O. stricta*, and the narrow-jointed
O. aurantiaca.

At the same time some of the specific combinations of
insect and host, and of different insects attacking the one
host, are new. The most extreme example of this is the

TABLE I. *Species of Cactaceae naturalized in Australia, their origin, and their current distribution within Australia. Based on Britton and Rose (1919), Dodd (1940), Mann (1970), and Willis (1972).*

Species	Native to[a]	Australian Distribution[b]
TRIBE OPUNTIEAE		
Nopalea cochenillifera (L.) Salm-Dyck	C.Am,WI	Qld
N. dejecta Salm-Dyck	Pan	Qld
Opuntia (subgenus Cylindropuntia)		
[c]*O. cylindrica DC.*	Peru,Ec	Vic
[c]*O. imbricata (Haw.) DC.*	Mex,USA	NSW,Qld,SA
O. subulata Engelm.	Ch	Qld
Opuntia (subgenus Platyopuntia)		
O. amyclea Tenore	Mex	Qld
[c]*O. aurantiaca Lindl.*	Ur	NSW,Qld
O. compressa (Salisbury) Macbride	USA	NSW
O. dillenii (Ker-Gawler) Haw.	C.Am,Mex,USA,WI	Qld,SA
O. elatior Mill.	Col,Pan,V	NSW,Qld
O. ficus-indica (L.) Mill.	C.Am	NSW,Qld,Vic
O. lindheimeri Engelm.	USA	SA
O. megacantha Salm-Dyck	Mex	NSW,Qld
O. microdasys (Lehmann) Pfeiff.	Mex	NSW
O. pachona Griffiths	Mex	NSW
[c]*O. robusta Wendl.*	Mex	Vic
O. rufida Engelm.	Mex	SA
O. sp. ("Joconoxtle")	Mex	Qld
[c]*O. streptacantha Lem.*	Mex	NSW,Qld
[c]*O. stricta Haw. (incl. O. inermis DC.)*	Cuba,USA	NSW,Qld,SA,Vic
O. sulphurea G.Don in Loud.	Arg,Ch	Qld
[c]*O. tomentosa Salm-Dyck*	Mex	Qld,SA
[c]*O. vulgaris Mill. (=O.monacantha Haw.)*	Arg,Br,Par,Ur	NSW,Qld,SA,Vic,WA
TRIBE CEREEAE		
[c]*Acanthocereus pentagonus (L.) Britt. & Rose*	C.Am,Mex,USA,WI	Qld
Cleistocactus baumannii Lem.	Arg,Par,Ur	Qld
Echinopsis multiplex (Pfeiff.) Zucc.	Br	Qld
[c]*Eriocereus martinii (Lab.) Ricc.*	Arg,Par	Qld
E. regelii Weing.	Arg (?)	Qld
E. fortuosus Forbes	Arg	Qld
Nyctocereus serpentinus (Lagasca & Rodrig.) Britt. & Rose	Mex	Qld

a Arg = Argentina, Br = Brazil, C.Am = Central America, Ch = Chile, Col = Colombia, Ec = Ecuador, Mex = Mexico, Pan = Panama, Par = Paraguay, Ur= Uruguay, V = Venezuela, WI = West Indies.

b NSW = New South Wales, Qld = Queensland, SA = South Australia, Vic = Victoria, WA = Western Australia.

c Pest species.

archetypal biological control combination: *O. stricta* from North America and *C. cactorum* from Argentina. *O. stricta* is also commonly attacked by strains of the scale *Dactylopius opuntiae*. These strains were derived from *O. lindheimeri* in Texas and *O. phaeacantha* Engelm. var. *discata* (Griffiths) Benson and Walkington in Arizona, yet *D. opuntiae* does not

TABLE II.   *Cactus-feeding insects successfully established in Australia, their origin, and their native and Australian host-plants. Based on Dodd (1940), Mann (1970), McFadyen and Tomley (1981)*

Species	Native to[a]	Host-plants Native	Host-plants Australian
LEPIDOPTERA			
PHYCITIDAE			
Cactoblastis cactorum (Berg)	Arg,Par,Ur	Tree, shrub and narrow-jointed platyopuntias, Cleistocactus sp.	O. aurantica, O. stricta, O. streptacantha, O. tomentosa
Tucumania tapiacola Dyar	Arg	Narrow-jointed platyopuntias	O. aurantiaca
COLEOPTERA			
CERAMBYCIDAE			
[b]Moneilema ulkei Horn	Mex,USA	Platyopuntias and cylindropuntias	O. stricta, O. streptacantha
[b]Moneilema variolare Thomson	Mex	Platyopuntias and cylindropuntias	O. stricta
[b]Archlagocheirus funestus Thomson	Mex	Tree platyopuntias	O. tomentosa, O. streptacantha
Alcidion cereicola Fisher	Arg,Par	Cereeae	E. martinii
CURCULIONIDAE			
Eriocereophaga humeridens O'Brien	Br	Cereeae	E. martinii
HEMIPTERA			
COREIDAE			
Chelinidea tabulata (Burmeister)	C.Am,USA	Platyopuntias	Platyopuntias, E. martinii
ERIOCOCCIDAE			
Dactylopius opuntiae Lichtenstein (various strains)	Mex,USA	Platyopuntias	Platyopuntias, N. dejecta
D. sp. nr confusus Cockerell	Arg	Narrow-jointed platyopuntias	O. aurantiaca
D. ceylonicus Green	Arg,Br,Ur	Tree and narrow-jointed platyopuntias	O. vulgaris
D. newsteadii Cockerell	Mex,USA	Cylindropuntias	O. imbricata
PSEUDOCOCCIDAE			
Hypogeococcus festerianus (Lizer y Trelles)	Arg	Cereeae	E. martinii, E. tortuosus

a Abbreviations as in Table I.
b Possibly established.

attack *O. stricta* where their ranges overlap in Texas (Mann, 1970).

This suite of established species makes it theoretically possible to study the evolution of organism-environment and organism-organism interactions. Given knowledge of the ecology and genetics of the organisms in their native habitats it should be possible to determine rates of divergence and adaptation applying to many aspects of ecology. Unfortunately such ecogenetic knowledge is limited. For example, in spite of detailed work on *Drosophila buzzatii*, both in Australia (Barker and Mulley, 1976; Mulley, James and Barker, 1979; Barker, 1982) and elsewhere (Fontdevila *et al.*, 1981; Sene *et al.*, 1982), its ecology in South America is still poorly known. Barker and Mulley (1976) noted a strong dependence of Australian *D. buzzatii* on the rot pockets generated in *Opuntia* spp. by *Cactoblastis* attack, but no such herbivore involvement has been reported elsewhere for this or for other cactus-*Drosophila* interactions. Nevertheless phycitid moths, especially *Melitara*, *Olycella*, and *Cactoblastis*, are widespread specialist herbivores of cacti in both North and South America (Mann, 1969), so it seems unlikely that the Australian association is unique. Their role in generating cactus rots deserves more attention.

III *Opuntia* Variation and *Cactoblastis* Population Dynamics

Cactoblastis cactorum survives today in the common shrub pest-pear *O. stricta*, the tree-pears *O. tomentosa* and *O. streptacantha*, and the narrow-jointed tiger pear *O. aurantiaca*. Susceptibility to attack is determined by the egg-laying behaviour of the female moths, which fly close to the ground, so that shrub-pear is readily found. Consequently only young seedling tree-pears or the inedible woody trunks of older specimens are located. Tiger pear is readily attacked but is not common through most of the range of *O. stricta*. Thus in most areas *C. cactorum* is concentrated on *O. stricta*. Dodd (1940) reports experiments on egg-laying preferences of caged moths for isolated cactus segments. These showed that *O. tomentosa* was avoided by *Cactoblastis*. This correlates with the observation (Dodd, 1940; Monro, 1975; Murray, unpublished) that in southern Queensland, in mixed populations of *O. stricta* and *O. tomentosa* even seedlings of *O. tomentosa* are rarely attacked. In spite of this there are large areas in the Gilbert and Expedition Ranges of central Queensland where shrub-pear is virtually absent and where *O. tomentosa* now shows heavy infestation. Therefore while it is possible that *C. cactorum* is in the process of adapting more intimately to *O. tomentosa*, ecological studies to date have concentrated on *O. stricta*, and these currently provide the basis for our understanding of the dynamics of Australian cactus ecosystems. Australian authors generally distinguish between two forms of shrub-pear: *O. stricta*, the spiny pest pear, and *O. inermis*, the smooth pest-pear (Dodd, 1940; Mann, 1970).

American workers on the other hand regard these as synonymous and referrable to *O. stricta* (Britton and Rose, 1919; Benson, 1962, 1969; Weniger, 1970; Bravo, 1978) on the basis that variations in spininess in mainland and West Indies populations are continuous, and are known to be strongly affected by environmental factors such as shading and cultivation. Moreover, Benson (1962, Fig. 3-19, 1969) and Bravo (1978) include the even spinier *O. dillenii* (Ker-Gawler) Haw. as a variety of *O. stricta* because of intergradations observed in Florida. Both Dodd (1940) and Mann (1970) justify the continued use of *O. inermis* for the smooth pest-pear but note that it should more properly be called *O. bentonii* Griffiths. Unfortunately there are serious problems with this usage: Britton and Rose (1919), Benson (1969) and Bravo (1978) all include *bentonii* as a synonym of *stricta*, while Weniger (1970) describes it as even *spinier* than *stricta*, and refers it to *O. engelmannii* Salm-Dyck var. *alta* Griffiths. The attempt to retain *stricta* as a name for the spinier Australian form is equally perverse: the original description (translated in Mann, 1970) is of a *smooth* pear - the "numerous" spines being very short, and obviously corresponding to glochids rather than large spines in the sense of later morphologists.

There is a clear biological need to distinguish the two major forms within Australia, because they are, in some areas at least, morphologically distinct. Moreover, they differ in their susceptibility to different strains of *Dactylopius opuntiae* (Mann, 1970), and variations in spininess have been correlated with *C. cactorum* egg-laying preferences (Dodd, 1940; Murray and Monro, unpublished). The source of the problem undoubtedly lies in the breeding system: apomixis is common in *Opuntia*, and its susceptibility to biological control measures has been attributed to limited sexual recombination (Burdon *et al.*, 1981). It seems likely that *O. stricta* in Florida and Texas comprises a spectrum of genotypes, of which two principal ones have been introduced to Australia, where they have maintained their integrity through clonal propagation and predominantly apomictic seed reproduction. Electrophoretic studies are in progress to test this hypothesis and should provide a more suitable framework for description.

These and other variations in *O. stricta* phenotypes have been implicated in the maintenance of stability of the *Opuntia-Cactoblastis* ecosystem. Dodd (1940) has described the dramatic initial impact of *C. cactorum* on *O. stricta*. Since then, both plant and herbivore have continued to co-exist at low and more or less stable levels, in spite of the moth's capacity to eliminate far more of the plant than it does.

Andrewartha and Birch (1954) and Nicholson (1957) accounted
for this co-existence on the basis of a "hide-and-seek" model
of local population extinction and recovery. However, it was
shown by Monro (1967) that the stability is not based on such
a process, but on the moth's behaviour of over-loading particular plants with eggs (laid in aggregates known as egg-sticks),
resulting in the starvation of over-crowded larvae. Whereas
Monro (1967) interpreted this behaviour as a population
regulation mechanism, it can also be argued that the stability
is an incidental outcome of selection for some other aspect
of egg-laying behaviour. Accordingly, the correlations
between plant attributes and egg-laying attractiveness were
investigated by Myers *et al.* (1981). They showed that under
natural conditions variation in attractiveness can be attributed to size of plant, proximity to prior attack, and colour
of plant. The first two relate to the apparency of the plant
to searching moths, but laboratory experiments suggest that
the moths exercise choice for colour. However, because the
plants chosen are generally greener, with higher nitrogen
levels, the behaviour is one that usually maximizes larval
survival and hence individual fitness, so that any higher-
order selection for stability *per se* seems unlikely.

The above characteristics identified by Myers *et al.*
(1981) are largely environmentally determined and the studies
were carried out in populations of predominantly smooth pest-
pear. Other studies have also noted variation in resistance
to attack amongst different phenotypes of *O. stricta,* and
whilst some forms are probably "resistant" through environmentally determined low moisture or nutrient levels (Dodd,
1940; White, 1981), others are not necessarily so. For
example, when smooth and spiny plants grow inter-mixed, laying
preferences for spiny plants can be found (Monro and Murray,
unpublished data). This result accords with Dodd's (1940)
laboratory tests on laying preferences.

In summary, there is much phenotypic variation within
Australian populations of *O. stricta,* and some of this is
genetically determined. How much variation, and how this
variation relates to the plant-choice behaviour of egg-laying
C. cactorum needs to be explored. Understanding this is
crucial because this behaviour determines the population
dynamics of the two species, and, secondarily, it determines
the environments of stem-rot organisms dependent on such
attacks. Parallel studies of these interactions between plant
phenotype, attacks by species of cactus-feeding moths, and
successionally dependent organisms need to be carried out in
North and South America as well as in Australia.

IV Evolution in Introduced Populations of *C. cactorum*

The circumstances of its introduction and the history of its spread combine to promote *C. cactorum* as an organism of considerable interest for evolutionary genetics. All Australian populations are derived from one initial introduction of 2,750 eggs in 1925. Within five years of release, the density had risen to 2.5×10^7 larvae per hectare, over the thousands of square kilometres of dense prickly pear in southern Queensland (Dodd, 1940). In the 57 years (114 generations) since its introduction, the moth has come to occupy the geographic range shown in Figure 1. Although humans

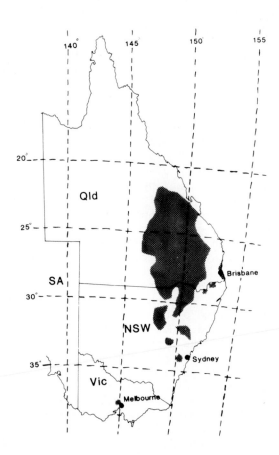

FIGURE 1. *Distribution of Cactoblastis cactorum in Australia (shaded areas).*

encouraged its initial spread, natural dispersal was also
considerable and by 1931 the moth had more or less occupied
its present range (Dodd, 1940).

Possible geographical differentiation in life history
characteristics was brought to light by the discovery in
1979 of an isolated population of *C. cactorum* living in *O.
stricta* at Bulla, near Melbourne, Victoria (lat. 38°S). The
nearest known population is at Camden, near Sydney (Lat.
34°S) 700 km away. Enquiries from landowners and government
weed control operatives disclosed that *C. cactorum* has been
present at Bulla since at least 1955 and possibly since 1945.
This is surprising since later attempts to introduce
C. cactorum to Victoria have failed (Parsons, 1973),
apparently due to inability of small larvae to survive
autumn and winter frosts. Subsequent studies have revealed
that this population of the moth is univoltine. This con-
trasts with the bivoltine pattern found elsewhere in
Australia (Figure 2) and in the Argentine source populations.
The life-history shift allows the moth to enter winter as a
large larva, the most cold-tolerant phase of the cycle.

While there is evidence of univoltinism for *Cactoblastis*
at the southern limit of the American distribution in
Uruguay (Lat. 35°S), the failure of other attempted Victorian

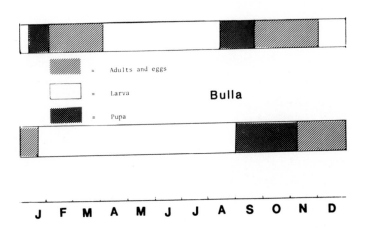

FIGURE 2. *Life-cycles of C. cactorum in Queensland and
New South Wales and at Bulla, Victoria.*

introductions suggests that the alteration is not simply environmentally determined. Detailed comparisons of life history traits and of adult morphology are currently being made between *C. cactorum* populations from throughout south-eastern Australia. Results available for numbers of days from egg-laying to hatching (incubation times) are shown in Table III. At a constant 18°C there is no difference between eggs laid by moths from Queensland and those from Bulla, but Hunter Valley eggs develop faster. However in a cyclic 12 hours dark/12 hours light, 11°C/28°C environment, where development is more rapid for both Bulla and Hunter Valley eggs, some Bulla eggs hatch extremely quickly. In the field, rapidly-developing Bulla eggs hatch in less than half the time of any recorded by Dodd (1940). While these results cannot yet be put into their full ecological context they show that in spite of deriving from a single initial sample, populations of *C. cactorum* have diverged substantially in the relatively short period of about 100 generations.

Further evidence of differentiation has come from electrophoretic studies. Murray, Adams and Baverstock (unpublished) have examined samples of 20 larvae from each of two Queensland populations and from Bulla. Populations are highly variable: 16 of 24 scorable loci are polymorphic in Queensland and 13 out of 23 at Bulla. The average heterozygosity per locus per individual is 0.28. At Bulla some rare alleles were not detected, and there are significant

TABLE III.   Incubation Times (Days) for Eggs from Different Populations of C. cactorum

	South Qld.	Hunter Valley, NSW	Bulla, Vic.
Mean (st. error) 18°C	63.9(1.9)	46.9(2.5)	62.3(0.6)
Mean (st. error) 11°/28°C	–	35.2(1.5)	41.1(1.5)
Minimum 11°/28°C	–	27	14
Minimum Field	18[a]	23[a]	7

[a]*From Dodd (1940)*

differences from the Queensland sample in the allele frequencies at four loci.

As well as revealing differences between Australian populations of *C. cactorum*, these results focus attention on possible causes of such a high level of variation. Studies on *Cactoblastis* and related genera, in both Australia and the Americas, should be profitable in the search for features of the cactus environment contributing to the maintenance of polymorphism, an area of interest to workers on other cactophilic insects (Barker, 1982).

There is a second reason for interest in the high level of variability in *C. cactorum*. The Australian populations are derived from Argentina through a simple bottleneck and flush sequence of population history. Dodd (1940) records that Australian moths are derived from a sample of 2,750 eggs. He also notes that the average egg production of the subsequent generation of females was a very high 191, whilst the lowest for most Australian *C. cactorum* is 50. Using these figures as extremes the original sample of eggs would represent the progeny of between 14.4 and 55 females, or approximately 29-110 genomes. While this is not a sufficiently small sample to affect seriously most levels of genetic variation (Nei *et al.*, 1975; Myers and Sabath, 1981), the original collections of *C. cactorum* were of *families* of larvae and they were all from the one locality, so that the effective population size may have been lower. This, taken together with the subsequent population flush and the evidence for rapid differentiation between Australian populations, makes *C. cactorum* a candidate for genetic transilience, as described by Templeton (1980). Whether or not this is the case can only be decided by looking at the genetic structure of Argentine populations and comparing them with Australian ones.

Additional opportunities to document rates of evolution and to examine the effects of bottlenecks are provided by the spread of *C. cactorum* for the control of *Opuntia* spp. in other parts of the world. Table IV shows that all subsequent introductions have been derived either directly or indirectly from the Australian populations, so that a sequence of population divergences can be established. While the numbers involved in most of these later introductions are large, there are two cases which are more interesting. In both Hawaii and the West Indies, deliberate introductions have been followed by rapid natural spread between islands (García Tuduri and Martorell, 1971). Moths introduced to Nevis in 1957 and to Montserrat in 1960 had reached St. Croix in the US Virgin Islands by 1963 and were extensively established

TABLE IV.   *Introductions of C. cactorum for Control of*
Opuntia spp.   Data from Dodd (1940),
Fullaway (1954), Holloway (1964), Simmonds
and Bennett (1966)

Introduced to	From	Year	Number of eggs introduced
Australia	Argentine	1925	2,750
South Africa	Australia	1927	12,000
		1932	112,600
New Caledonia	Australia	1932	10,000
		1933	25,000
Hawaii[a]	Australia	1950	205,200
Mauritius	South Africa	1950	-
Leeward Is.[a]	South Africa	1957	15,000[b] (+100 larvae)

[a] *Followed by natural spread to other islands (see text)*
[b] *Counting one egg-stick as 50 eggs.*

along the southern coast of Puerto Rico in 1966, where they
were attacking *O. dillenii*. This natural colonization needs
to be genetically documented, preferably before *C. cactorum*
reaches Florida and begins attacking *O. stricta* on its home
ground.

It should be stressed that an evolutionary approach to
biological control introductions is possible for many other
organisms. Easteal (1981) has developed a similar model for
the diversification of the marine toad *Bufo marinus*. In the
sphere of cactus insects species of *Dactylopius* offer excellent material, as does the most recent successful establishment in Australia,   the Argentine mealybug *Hypogeococcus
festerianus*, which was introduced for the control of
Eriocereus spp. in 1975 (McFadyen and Tomley, 1981). The
stock released was based on the progeny of only six females,
and the improving rate of attack of the cactus in the field
suggests that some alterations to the insect may already have
occurred (W. Haseler, pers. comm.).

V Conclusions

The cacti and associated insects introduced to Australia provide combinations of species which are not found elsewhere. One such combination, the interaction between *O. stricta* and *C. cactorum*, provides the basic environment on which other cactus-specific organisms depend. Studies on these species in Australia have revealed our ignorance of their biology in their native conditions. In particular, the high genetic variability of *C. cactorum* suggests that it and other cactus-breeding phycitids would be valuable material for workers seeking to understand variation in other cactophilic insects. As well, adaptive diversification appears to have occurred within Australian *C. cactorum* and attention is drawn to the potential of the moth's rapid spread elsewhere in the world for providing an extensive experiment in evolution.

REFERENCES

Andrewartha, H.G., and Birch, L.C. (1954). "The Distribution and Abundance of Animals." University of Chicago Press, Chicago.

Barker, J.S.F. (1982). Chapter 14, this Volume.

Barker, J.S.F., and Mulley, J.C. (1976). *Evolution 30*, 213.

Benson, L. (1962). "Plant Taxonomy." Ronald Press, New York.

Benson, L. (1969). *In* "Flora of Texas," Vol. 2 (C.L. Lundell, ed.), p. 221. Texas Research Foundation, Renner, Texas.

Bravo, H.H. (1978). "Las Cactáceas de México," Vol. 1. Universidad Nacional Autónoma de México, México.

Britton, N.L., and Rose, J.N. (1919-24). "The Cactaceae," 4 vols., Carnegie Institute of Washington, Publ. 248, Washington.

Burdon, J.J., Marshall, D.R., and Groves, R.H. (1981). *Proc. V Int. Symp. Biol. Contr. Weeds*, 21.

Dodd, A.P. (1940). "The Biological Campaign against Prickly Pear." Govt. Printer, Brisbane.

Easteal, S. (1981). *Biol. J. Linn. Soc. 16*, 115.

Fontdevila, A., Ruiz, A., Alonso, G., and Ocaña, J. (1981). *Evolution 35*, 148.

Fullaway, D.T. (1954). *J. Econ. Entomol. 47*, 696.

Garcia Tuduri, J.C., and Martorell, L.F. (1971). *J. Agr. Univ. Puerto Rico 55*, 130.

Holloway, J.K. (1964). *In* "Biological Control of Insect Pests and Weeds" (P. de Bach, ed.), p. 650. Chapman and Hall, London.

Mann, J. (1969). "Cactus-Feeding Insects and Mites." Smithsonian Institution Press, Washington.

Mann, J. (1970). "Cacti Naturalised in Australia and Their Control." Govt. Printer, Brisbane.

McFadyen, R.E., and Tomley, A.J. (1981). *Proc. V Int. Symp. Biol. Contr. Weeds,* 589.

Monro, J. (1967). *J. Anim. Ecol. 36,* 531.

Monro, J. (1975). *Proc. Ecol. Soc. Austr. 9,* 204.

Myers, J.H. (1978). *Proc. IV Int. Symp. Biol. Contr. Weeds,* 181.

Myers, J.H., and Sabath, M.D. (1981). *Proc. V. Int. Symp. Biol. Contr. Weeds,* 91.

Myers, J.H., Monro, J., and Murray, N.D. (1981). *Oecologia 51,* 7.

Nei, M., Maruyama, T., and Chakraborty, R. (1975). *Evolution 29,* 1.

Nicholson, A.J. (1957). *Ann. Rev. Entomol. 3,* 107.

Osmond, C.B., and Monro, J. (1981). *In* "Plants and Man in Australia" (D.J. and S.G.M. Carr, eds.), p. 194. Academic Press Australia, Sydney.

Parsons, W.F. (1973). "Noxious Weeds of Victoria." Inkata Press, Melbourne.

Sene, F.de M., Pereira, M.A.Q.R., and Vilela, C.R. (1982). Chapter 7, this Volume.

Simmonds, F.J., and Bennett, F.D. (1966). *Entomophaga 11,* 183.

Templeton, A.R. (1980). *Genetics 94,* 1011.

Waddington, C.H. (1965). *In* "The Genetics of Colonizing Species" (H.G. Baker and G.L. Stebbins, eds.), p. 1. Academic Press, New York.

Weniger, D. (1970). "Cacti of the South-West : Texas, New Mexico, Oklahoma, Arkansas and Louisiana." University of Texas Press, Austin.

White, G.G. (1981). *Proc. V Int. Symp. Biol. Contr. Weeds,* 609.

Willis, J.H. (1972). "A Handbook to Plants in Victoria, Vol. II Dicotyledons." Melbourne University Press, Melbourne.

PART II
DROSOPHILA PHYLOGENY
AND SYSTEMATICS

3
Pathways of Evolution in the Genus *Drosophila* and the Founding of the *Repleta* Group[1]

Lynn H. Throckmorton

Department of Biology
University of Chicago
Chicago, Illinois

I Introduction

When the biogeography of the *repleta* group was last treated (Throckmorton, 1975), it was not possible to present either the evidence for, or the theoretical basis of, conclusions that were reached. For a group being so actively investigated, it is desirable that this information be available, but space still limits what can be provided. Only token examples of the data can be given here, but the method of biogeographical analysis is summarized. In addition, some new evidence is in hand, justifying a slightly different perspective on the *repleta* group and its nearest relatives from what was seen before. The results of this reinterpretation are also presented.

The discussion is organized according to the following plan. *First*, requirements for biogeographical analysis are sketched out, indicating what is needed before a large-scale biogeographical study is begun. *Second*, phylogenetic relationships among the major groups of the Drosophilidae are shown, together with the kind of evidence from which they were inferred, and the method used for phylogenetic analysis is indicated briefly. *Third*, the method of biogeographical analysis is given in some detail, with small extracts from the data illustrating the kind of information that supports

[1]*This work was supported in part by grants GM 23007 from the National Institutes of Health and AG 01941 from the National Institute of Ageing.*

ECOLOGICAL GENETICS AND EVOLUTION
ISBN 0 12 078820 9

the most important biogeographical conclusions. *Fourth*, data from paleoecology is integrated with ecological data from extant species to show how observed biogeographical patterns justify conclusions regarding the time and place of origin of the *repleta* group. And, finally, brief note is taken of the information still needed to further refine our understanding of the group's origin and affinities.

II Requirements for Biogeographical Analysis

More and more, it is apparent that evolutionary study cannot be provincial, and this is especially true for studies of systematics and biogeography. Evolutionary changes in a group may be largely conditioned by properties of the lineage from which it sprang, distribution patterns on one continent may contradict rather than complement those on another, and patterns that seem simple when viewed from local perspective may appear quite otherwise, and much more complex, when seen in world view. Regional studies may not even be able to treat with species authoritatively, much less with species complexes or species groups. Thus it is, that biogeographical analysis, even of the compact and geographically coherent *repleta* group, must encompass most of the family Drosophilidae and much of its distribution throughout the world.

There are four requirements for a thoroughgoing biogeographical analysis: (1) a phylogeny, derived from intrinsic properties of individuals from the entire group which *includes* the group of primary interest, (2) knowledge of the distribution of all species, species complexes, and species groups within this larger group, (3) knowledge of the ecological requirements of all species of the larger group, and (4) knowledge of the paleoecology of the regions throughout which the presentday species are distributed and through which their ancestral forms may have dispersed.

This is a formidable list of requirements, and it is rarely fulfilled. At present, knowledge of drosophilid distribution and ecology is poor, our knowledge of seasonal and altitudinal relationships within faunas is virtually nonexistant, and the basic taxonomic treatment of major faunas is still very incomplete (e.g., Rocha Pité and Tsacas, 1979). Paleoecology itself will probably never be as informative as hope would have it, and historical detail on past plant associations, their distributions, and their spatial and temporal changes, will surely come slowly. Answers are, accord-

ingly, limited by available data.   They will change as the
data change.   On a positive note, however, there is reason
to expect that future revisions of drosophilid biogeography
will not be extreme.   Major new discoveries from Africa
(Tsacas, 1979, 1980) and Australia (Bock, 1976, 1977, 1979,
1980a; Bock and Parsons, 1975, 1978) provide data complement-
ing and reinforcing patterns seen earlier.   This important
new information adds tremendously to our understanding of
drosophilid evolution, but it shifts the biogeographical
pattern hardly at all.

III   Phylogenetic Relationships

A phylogeny is already available for the major genera
and subgenera of the Drosophilidae (Throckmorton, 1962, 1965,
1966, 1975).   The evidence upon which it is based comes from
attributes of a variety of anatomical features of the eggs,
pupae, and adults, and from both males and females.   Data are
derived from more than 40 genera and subgenera of drosophilids,
and from more than 50 species groups, just of the genus
Drosophila.   Nearly two-thirds of the data are published
(Throckmorton, 1962, 1966), and significant unpublished information is presented herein.

Figure 1 shows the overall structure of the family,
indicating the relative positions of major genera, subgenera,
and so on.   The central problem in phylogenetic analysis is
always determining direction of evolution, and this can be
done objectively only by reference to "outside" groups
(Throckmorton, 1962, 1968).   My earliest publications on the
phylogeny of *Drosophila* (Throckmorton, 1962, 1965) were
limited because too few data from the earliest radiations in
the family were available.   Other families of diptera were
used as outside groups, and they turned out to be somewhat
too distant.   Since then, I have accumulated data on an as-
sortment of steganine genera, and this has allowed correction
of some early mistakes (compare Throckmorton, 1965, 1975).
It also confirmed the placement of the Steganinae as an early
radiation of the family, out of which sprung the Drosophi-
linae, the genus *Drosophila*, and ultimately, the *repleta*
group.

Figures 2 and 3 illustrate some of the features contributing to this conclusion.   They are intended, first to
show the relatively pronounced anatomical change that has
occurred during the evolution of the family, which readily

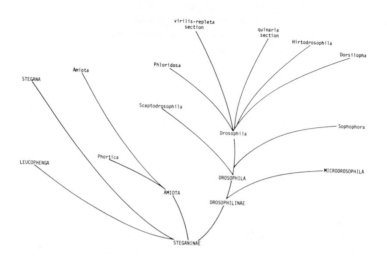

FIGURE 1. Phylogenetic relationships of major groups within the family Drosophilidae. Names in upper-case letters, other than those of subfamilies, are of genera. Only the first letter is capitalized for names of subgenera.

permits assessing direction of evolution; and second, taking together the features shown in both figures, to illustrate covariation of character states, which is the analytical evidence for phylogeny (Throckmorton, 1978). Thus, the uniqueness of particular character states among *Drosophila* is readily seen by contrast to states among the Steganinae. And, if space permitted, the uniqueness of steganine character states could be seen in contrast to states among other families of diptera. When unique states of different characters tend to "agree" on specific partitions among groups, as they do in the examples shown, clusters of species appear, and these are inferred to derive from a unique common ancestor. It will be recognized, of course, that the branching pattern in the figures is derived from consideration of the complete data set, and not just of the features illustrated.

For the phylogeny of the family, this brief summary must suffice. It shows the kind, but not the extent, of the basic data. Only key features of phylogenetic analysis are given, since method is treated extensively elsewhere (Throckmorton, 1962, 1965, 1968, 1978).

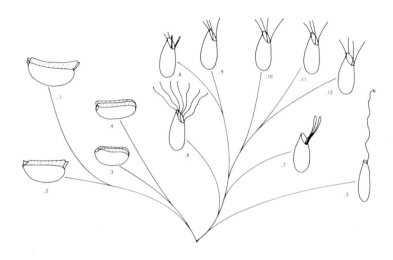

FIGURE 2. *Eggs of species from the taxa shown in Figure 1. Names are as follows: .1 Stegana nigrifrons, .2 Leucophenga ornata, .3 Amiota (Phortica) variagata, .4 Amiota (Amiota) clavata, .5 Microdrosophila urashimae, .6 Drosophila (Scaptodrosophila) nitithorax, .7 D. (Sophophora) simulans, .8 Dettopsomyia nigrovittata, .9 D. (Drosophila) aldrichi, .10 D. (Drosophila) cardinoides, .11 D. (Hirtodrosophila) pictiventris, .12 D. (Drosophila) busckii.*

IV Biogeography

Once a phylogeny is available, the basic procedures of biogeographical analysis are simple. One lineage at a time, known distributions of major species complexes are plotted on the world map. Working back from derivative groups, connections are drawn between the geographical areas they occupy until the base of a lineage is reached. When all individual lineages are mapped, the bases of lineages are then connected to each other in accordance with their phylogenetic relationships. When the bases of phylogenetically close lineages plot into the same geographical region, the lineages are presumed to have arisen there in the temporal sequence indicated by the phylogeny. When the bases of phylogenetically close lineages plot into widely separate areas, it is necessary to infer the routes or means by which ancestral populations achieved this distribution, and how the disjunction was established. The combined information from distribution

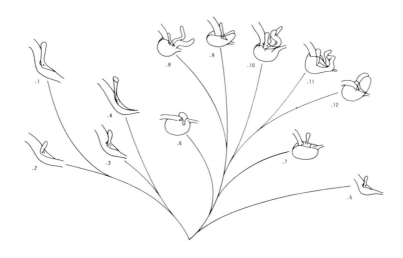

FIGURE 3.   Ejaculatory bulbs of species of representative genera and subgenera of drosophilids.   Names are as in
Figure 2.

and phylogeny produces biogeographical patterns.   To determine how and why these originated, recourse is made to the
known ecology of extant groups, and to paleoecology.   Ex-
planations that emerge may permit inferences regarding probable timing of events, the separation of certain lineages,
and so on.

The use of this method to discover the place of origin
of the genus *Drosophila* is illustrated in Figure 4.   Since
phylogenetic analysis has already shown that the subgenus
Scaptodrosophila represents the earliest radiation in the
genus, attention can be given only to it.   The derivative
ends of its known lineages are found in South America, in
North America, in temperate Asia, and in Australia and
Africa.   Types most nearly like the primitive for the sub-
family (Fig. 3.5) are mostly in Southeast Asia (Fig. 4.1),
although the type seen in Africa (Fig. 4.4) is only slightly
more derivative.   Types known from South America and Africa
differ sharply (compare Figs. 4.4 and 4.10), and the connect-
ing forms between them are seen on the northern continents
(Figs. 4.7-4.9).   They trace back through intermediate types
(Fig. 4.5) to primitives in tropical Asia, and the African
type seems to be derived along another route.   Hence, in its

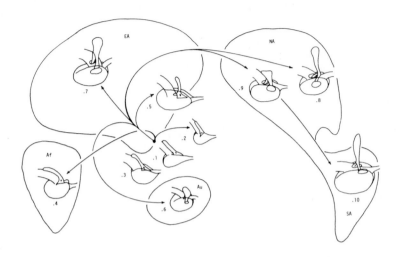

FIGURE 4. Geographical patterns illustrating evidence that the genus Drosophila originated in Southeast Asia. These examples are of ejaculatory bulbs from species of the subgenus Scaptodrosophila. Species names are as follows: .1 dorsocentralis, .2 coracina, .3 bryani, .4 latifasciae- formis, .5 subtilis, .6 lativittata, .7 pattersoni, .8 casteeli, .9 victoria, .10 D. sp. of the victoria group.

evolution, the subgenus *Scaptodrosophila* appears to have radiated outward, along several lines, from Southeast Asia. Again, it must be emphasized that this conclusion rests on other features also, but the ejaculatory bulbs show the pattern in its most diagrammatic state and so have been chosen to illustrate here.

Figure 5 illustrates the basic biogeographical pattern seen at least five times within the genus *Drosophila*. It shows examples from the sophophoran radiation, which follows next after *Scaptodrosophila* in the evolution of the genus. Ejaculatory bulbs are used again since it is easy to see, both their resemblances to primitive types, and their ad- vances relative to them. From this radiation, the *melano- gaster* group dominates tropical Asia and Africa, and it has made its way to Australia (Bock, 1980b). In Central and South America, this radiation is represented by the *saltans* and *willistoni* groups. It can be noted that in both the New and Old World tropics, the sophophoran radiation began from

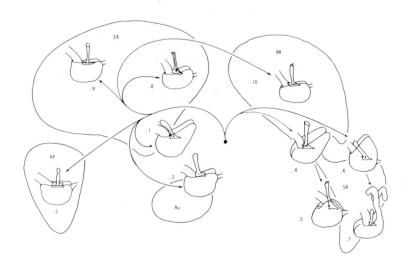

FIGURE 5.   *The basic biogeographical pattern seen in the genus Drosophila.   It is illustrated here with ejaculatory bulbs from species in the subgenus Sophophora.   Species names are as follows: (melanogaster gp.) .1 ananassae, .2 pseudo-takahashii, .3 melanogaster; (saltans gp.) .4 emarginata, .5 prosaltans; (willistoni gp.) .6 willistoni, .7 capricorni; (obscura gp.) .8 bifasciata, .9 subobscura, .10 algonquin.*

basically similar types (compare Figs. 5.1, 5.4 and 5.6), which are themselves most similar to primitive types from the subgenus *Scaptodrosophila* (compare Fig. 5.1 and Figs. 4.1-4.5).   In contrast, the North American temperate forest members of this radiation share their characters, not with Central and South American forms, but with species from temperate Asia (Figs. 5.8-5.10).   And these species themselves show their closest ties with species of the *melanogaster* group, mainly from Asia and Africa (compare Figs. 5.8-5.10 and Figs. 5.2-5.3).   On the one hand, there is continuity through the temperate zone, and on the other, tropical disjunction, with the temperate and tropical forms intergrading in the Old, but not in the New, World.   And the temperate forms are derivative to at least some of the Old World tropical species. Which is to say, the sophophoran lineage became distributed throughout the tropics, split into New and Old world branches, and the *obscura* group of the north temperate zone originated from the Old World lineage.

When this pattern was first observed (Throckmorton, 1975), it was a temptation to attribute it to continental drift, as Tsacas (1979) has done. He proposes that *Sophophora* arose on the southern supercontinent prior to the separation of Africa and South America during Cretaceous times. However, when phylogeny is considered as well as distribution, both of *Sophophora* and of related groups, this conclusion is not supported by presently available evidence. It is the derivative members of the *melanogaster* group and its relatives that are found in Africa, and the *saltans-willistoni* lineage itself is derivative to other sophophorans. The Oriental region has forms showing characters which tend to intergrade with types from *Scaptodrosophila*, so it is here the group must be inferred to have originated. One can, of course, hypothesize that the group originated elsewhere, disappeared there, and primitive types remain only in South-east Asia. But one cannot contradict an observation (primitive groups exist in Southeast Asia) with a hypothesis (they might have existed at one time elsewhere). Hence, parsimoniously, *Sophophora* arose in the tropics of Asia.

A similar pattern of tropical disjunction and north temperate continuity is seen for all the major radiations of the higher *Drosophila* (Throckmorton, 1975). Of the groups illustrated in Figure 1, it is most evident for *Sophophora* and *Hirtodrosophila*, and for the *quinaria* section of the subgenus *Drosophila*. It is somewhat less clear for the *virilis-repleta* section, apparently because very few representatives of this radiation are yet recognized as such from tropical Asia. Until additional evidence forces reevaluation, this pattern is most readily interpreted within the context of events of the Cenozoic Era. Among all of the groups, tropical origin is always clearly evident. Support is not equally strong for an origin in the Old, as opposed to the New, World, and some of the higher radiations may have originated in the New World. This does not affect the major question, however, which is that of accounting for disjunctions in so many tropical radiations.

As is well-known now (e.g., Pearson, 1978), in the early Tertiary the climate was mild. About 60 million years ago the southern continents were still mostly separate from the northern ones. India was an island drifting toward eventual collision with Asia, Australia was drifting eastward and northward, and Southeast Asia was probably a mass of island arcs and continental fragments aggregated in a

very complex fashion. In North America the Rockies were not
yet high, tropical forest extended northward at least to mid-
continent, and wet subtropical or warm temperate forest extended through Alaska and into east Asia. The Oligocene saw
the continued rise of the Rocky Mountains, the Alps, and the
Himalayas and climates of the northern land masses cooled. A
relatively pronounced change occurred about 30 million years
ago (Wolfe, 1978). During this period a flora of arid lands
began to expand to occupy much of southwestern North America.
A relatively cool, mixed mesophytic forest replaced tropical
or subtropical vegetation in Beringia, inaugurating the disjunction of tropical forests that remains to the present day.
By middle Miocene, about 20 million years ago, a diversified
broad-leaved forest probably extended from Japan through
Alaska into Oregon. Mean temperatures in Beringia continued
to decline, and by late Miocene time (7-10 million years ago)
there was a disjunction of the mesophytic forest, and coni-
fers began to occupy the uplands near the Bering Land Bridge
(Graham, 1972; Daubenmire, 1978; Pearson, 1978; Wolfe, 1978).

V The Origin of the *Repleta* Group

Granting the inference that the genus originated in
tropical Asia, and given the evidence that all its major
radiations were founded in both the New and Old Worlds by
tropical lineages, it seems necessary to conclude that the
major lineages of *Drosophila* themselves arose before
tropical connections between the New and Old Worlds were
broken; i.e., before 30 million years ago, more or less. A
fossil is known from Baltic amber (Hennig, 1965) establishing that drosophilids were in European forests by about 40
million years ago. A fossil is also known from amber of
Chiapas, Mexico (Wheeler, 1963), so drosophilids were on the
North American continent by about 30 million years ago. Accordingly, when all available information is combined, it is
consistent with, and supports, the inference that the ancestral lineage of the *virilis-repleta* radiation in the New
World tropics was in place in tropical North America by about
30 million years ago. There is no ground even to speculate on
how much before then they appeared. A somewhat later time
might be permitted, depending on the ecological requirements
of the forms involved, but 20 million years ago (middle Miocene) is probably the latest time members of a major tropical
radiation might have passed between the northern continents,
and even this is quite improbable, given current views of
Tertiary climates (e.g., Pearson, 1978). Possible long-

distance dispersals could be considered, but since a corridor existed at a reasonable time, and since a number of lineages made the passage, that does not seem called for.

Figure 6 shows the major lineages of the *virilis-repleta* radiation in the New World. This has been changed from the earlier version (Throckmorton, 1975) by data obtained from *D. melanissima*, a species of the eastern deciduous forest of North America, and not available to me earlier. It proves to connect the *annulimana* group and the *robusta* group. The *melanica* group is shifted farther from the *robusta* group and into a closer relationship with the *nannoptera, bromeliae,* and *peruviana* groups, as shown in the figure. Since both the *annulimana* group and the *nannoptera* cluster are known only from the New World, this now places the origin of both the *melanica* and *robusta* groups there.

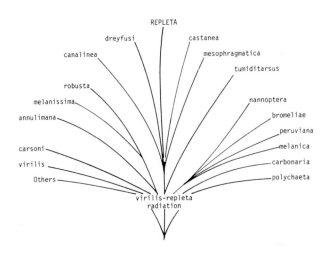

FIGURE 6. *The phylogenetic structure of the virilis-repleta radiation. The names are those of species groups. "Others" refers to species I have observed in the Orient but for which little information is available. They are clearly members of this radiation, but they have not yet been placed into species groups. Of the tropical or semitropical forms, only the polychaeta group is represented by both New and Old World species. However, polychaeta itself is nearly cosmopolitan and was probably introduced by man into the New World.*

So far as is known, only one Eurasian species, *D. tumidi-tarsus*, shows a possible relationship to the *repleta* group, but chromosomally it is not close (Wasserman, personal communication). Thus, the evidence indicates that the *repleta* lineage (including the *canalinea* group, etc.) originated from forms that were already established in the New World. This lineage may share a New World common ancestor with other New World lineages, but the present evidence from phylogeny does not require this (or exclude it). The *virilis-repleta* radiation in the New World began from species of the tropical forest. Whether the ancestor of the *repleta* group itself was a forest species which "became" a "repleta", began diversifying in the forest, and subsequently moved into arid habitats, or whether it first moved into arid habitats and became a repleta there, is difficult to determine. Its closest relatives, the *castanea, canalinea, dreyfusi* and *mesophragmatica* groups, are forest forms, for the most part, and apparently primitive members of the *repleta* group are at least facultative forest forms breeding in fallen fruit (see Throckmorton, 1975, p. 446). Parsimoniously, this permits the inference that the founder of the *repleta* group was a forest form, not necessarily of the wet forest, which "became" a repleta while still associated with forest habitats. The evidence for this, however, is not strong. To clarify the point further, it will be necessary to establish more firmly just which species are nearest to the bases of the different branches within the *repleta* group, and this will require much additional cytological and anatomical study. Then it will be necessary to determine the breeding sites, and the degree of polyphagy, of these species. This is also a formidible, but crucial, undertaking.

Determining the place of origin of the *repleta* group encounters somewhat the same problems as were met when considering the ecology of its founders. Earlier (Throckmorton, 1975), the known primitives from the major branches of the group showed distributions centering in and around Mexico That still has not changed, and, parsimoniously, the place of origin of the group must still be regarded as Mexico. However, if different species prove to be primitive, and if their distributions are different, this conclusion may change. At the present time, and on anatomical grounds especially, the major separation within the *repleta* group is between the *hydei* subgroup on the one hand and the remaining subgroups on the other, with the *fasciola* subgroup being most primitive among the latter forms. (For example,

see Throckmorton, 1962, and compare its Figures 17.1-17.8 to
Figures 4.1-4.4 of this paper.) Both the *canalinea* and *dreyfusi* groups appear closest to the *hydei* subgroup. The *castanea* and *mesophragmatica* groups are closest to each other
and somewhat more distant from the *repleta* group. Consequently the members of the *fasciola* subgroup take on unusual
importance with respect to the origin of the group, as do
members of the *hydei* subgroup and their relatives. Sene
(personal communication) has found a number of new *fasciola*
species from South America. Their features, when fully
known, may be decisive evidence for the location and ecology
of the founders of the *repleta* group, or at least for that
branch of the group that includes the *mulleri* complex and its
cytological derivatives. It is not inconceivable that the
repleta group had two (or more) "foundings", and the origin
of the *hydei* branch might profitably be treated as a separate
problem until it proves otherwise.

As was shown earlier, evidence indicates that the an-
cestors of the *repleta* group were in the New World by at
least 30 million years ago, and they almost surely arrived
there from Asia by way of Beringia. How long before then
they might have arrived cannot be said, and exactly when a
true "repleta" emerged from the ancestral radiation is
equally uncertain. There has clearly been much independent
evolution within the group on both the North and South American continents. The water gap between North and South
America existed from Eocene to Pliocene times (Pearson, 1978)
but it may not have been much of a barrier to *Drosophila*.
Arid conditions that might have provided habitats for an-
cestors of the *repleta* group were already present in northern
South America by the late Cretaceous (Pearson, 1978), and
they appeared in southwestern North America by Oligocene
times, or earlier (Daubenmire, 1978). Hence, by 30 million
years ago, the stage was surely set for the appearance of
the *repleta* group, if they had not already appeared by then.
Still, a great amount of evolution or diversification does
not seem to have occurred within the *repleta* group before 30
million years ago, or at least some of them should have "returned" to Asia through the same corridor that brought them
to North America. There is no evidence that that happened.
This could be explained as due to limitations on the move-
ments of their cactus hosts, but since some species of the
repleta group have been reared from bleeding trees, fallen
fruits and flowers (cf. Throckmorton, 1975, p. 446) this cannot be the whole story. Most probably, the major evolution

within the group has occurred during the last 30 million
years, especially during Miocene times when floras of arid
lands seem to have been well developed.

VI Further Needs

While a great deal can already be said about the origin
of the *repleta* group, much information is still needed at
almost all levels. The accuracy of conclusions relating to
this group depends heavily on the completeness of understanding of evolution of the genus as a whole. Refining our present knowledge requires more facts, especially about the
members of the *virilis-repleta* radiation, both in Asia and
in Africa. Almost surely, species exist in Southeast Asia,
knowledge of which would enable us to more sharply delineate
the evolutionary history of the entire radiation. And knowing that there are no close relatives of the *repleta* group
in Africa would reinforce present conclusions of the group's
origin on a northern continent. Timing of the origin of
the group depends surprisingly heavily on our knowledge, not
only of paleoecology, but also of the ecology of present
species. If the founder of the group were a high altitude
species, for example, the conditions for its passage to the
New World might be quite different from those envisioned by
the present treatment, which necessarily equates "tropical"
with "warm". Hence, major efforts are needed to obtain more
complete data on all attributes of species of the group.
Anatomical, cytological and ecological data all complement
each other, and they must be gathered, not just from
species of the *repleta* group, but from their close relatives
as well.

REFERENCES

Bock, I. R. (1976). *Aust. J. Zool., Suppl. Ser. No. 40.*
Bock, I. R. (1977). *Aust. J. Zool. 25*, 337.
Bock, I. R. (1979). *Aust. J. Zool. Suppl. Ser. No. 71.*
Bock, I. R. (1980a). *Aust. J. Zool. 28*, 261.
Bock, I. R. (1980b). *Systematic Entomology 51*, 341.
Bock, I. R., and Parsons, P. A. (1975). *Nature, Lond. 258*,
      602.
Bock, I. R., and Parsons, P. A. (1978). *Systematic Ento-
      mology 3*, 91.
Daubenmire, R. (1978). "Plant Geography." Academic Press,
      New York.

Graham, A. (1972). "Floristics and Paleofloristics of Asia
    and Eastern North America." Elsevier Publishing
    Company, Amsterdam.
Hennig, W. (1965). *Stuttg. Beitr. Naturkd. Nr. 145.*
Pearson, R. (1978). "Climate and Evolution." Academic
    Press, London.
Rocha Pité, M. T. and Tsacas, L. (1979). *Bolm Soc. port.
    Ciênc. nat. 19,* 37.
Throckmorton, L. H. (1962). *Univ. Texas Publ. 6205,* 207.
Throckmorton, L. H. (1965). *Syst. Zool. 14,* 221.
Throckmorton, L. H. (1966). *Univ. Texas Publ. 6615,* 335.
Throckmorton, L. H. (1968). *Syst. Zool. 17,* 355.
Throckmorton, L. H. (1975). *In* "Handbook of Genetics"   (R.
    C. King, ed.), p. 421. Plenum Press, New York.
Throckmorton, L. H. (1978). *In* "Beltsville Symposia in
    Agricultural Research. 2. Biosystematics in
    Agriculture." (J. A. Romberger, R. H. Foote, L.
    Knutson, and P.L. Lentz, eds.), p. 221. Allenheld,
    Osmun and Company, Montclair, N. Y.: John Wiley and
    Sons, New York.
Tsacas, L. (1979). *C. R. Soc. Biogéogr. 480,* 29.
Tsacas, L. (1980). *Annls. Soc. ent. Fr. (N.S.) 16,* 517.
Wheeler, M. R. (1963). *J. Paleontol. 37,* 123.
Wolfe, J. A. (1978). *Amer. Sci. 66,* 694.

4

Cytological Evolution in the
Drosophila repleta Species Group

Marvin Wasserman[1]

Biology Department
Queens College
Flushing, New York

I Introduction

The *Drosophila repleta* group consists of seventy-one described species and five, as yet, undescribed forms. Metaphase karyotypes and salivary gland chromosomes of sixty-two
of these species have been examined (see Wasserman, 1982 for
a recent review). The basic and most common metaphase karyo-
type consists of five pairs of autosomes (four pairs of rods
and one pair of dots), a rod-shaped X chromosome and a short
rod-shaped Y. These chromosomes tend to be small with little
to distinguish them. However, changes in the position of the
centromeres and also changes in the amount of heterochromatin
could be detected. The salivary gland chromosomes have been
particularly useful since they allow for a very precise and
accurate determination of chromosomal rearrangements. These
chromosomes are very large, multistranded with puffs, con-
strictions, and bands which characterize the individual
chromosomes and make even very small regions identifiable.
Moreover there is somatic pairing between homologous chromosomes so that even relatively small chromosomal mutations can
be detected in heterozygous individuals. By means of camera
lucida drawings and photographs, chromosomal maps are

[1]*Research supported by City University of New York PSC-
BHE Grant #12261 and N.S.F. Grant #DEB 79-21760*

ECOLOGICAL GENETICS AND EVOLUTION
ISBN 0 12 078820 9

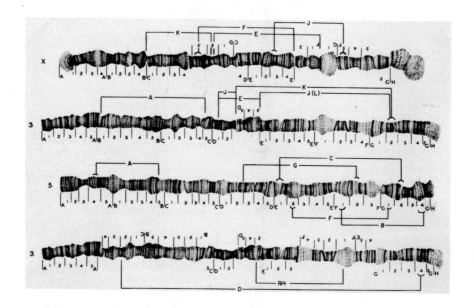

FIGURE 1.   *Salivary gland chromosomes of the mulleri subgroup.*

constructed upon which the specific cytological rearrangements can be delineated.   Figure 1 shows several typical chromosome maps.

TABLE I.   *Numbers of Chromosomal Mutations*

Type	Number
Deletions	0
Duplications	0
Reciprocal Translocations	0
Centric Fissions	0
Centric Fusions	4
Pericentric Inversions	1
Paracentric Inversions	235
Total	240

Table I lists the number of different types of chromosomal mutations which have been observed in the *repleta* group.
No reciprocal translocations, no duplications or deletions of
euchromatic material and no centric fissions have occurred.
Ward (1949) reported that Wharton found a strain of *bifurca*
with a pericentric inversion in the heterochromatin of the X.
Four centric fusions have occurred during the history of
these species. By far the most prevalent type of chromosomal
mutation to have survived has been the paracentric inversion
where 235 have been detected. These are particularly important because not only do they represent a character which
will be studied with a high degree of accuracy, but they are
unambiguous phylogenetic determinants. Sturtevant and
Dobzhansky (1936) presented arguments to show that overlapping inversions yield chromosomal phylogenies. For
example, given the gene orders: #1: ABCDEFGHIJ and #3:
ABFEHGCDIJ, one must propose an intermediate, hypothetical,
gene order #2: ABFEDCGHIJ, because it is impossible to convert either #1 directly to #3, or the reverse. Possible
phylogenies are then (A) 1→2→3; (B) 3→2→1; (C) 1←2→3.

The inversions, themselves, do not indicate direction.
However if, by using other criteria, primitiveness is established, direction of evolution can be determined. For
example, in Figure 1, inversion 5c and 5g overlap each other.
Since the 5 standard is the primitive, these three gene orders exemplify situation (C), where standard is #2 and 5c
and 5g are mutually exclusive divergent types. Non-
overlapping, independent inversions are also useful.
Wasserman (1963) presented arguments which showed that these
can also be used as phylogenetic determinants because each
inversion is a unique event. Therefore any two populations
or species which have the same inversion are more closely
related to each other than either is to a third population
or species which lacks the inversion.

The gene order found in *D. repleta* was considered the
standard for the whole *repleta* group. All other species
were compared back to that standard. Each inversion was
given a binary name: the first part indicating the chromosome (X,2,3,4,5) and the second indicating the specific inversion (a,b,c,etc.). Many inversions were found and super-
scripts were required to specify the inversions ($2d, 2d^2, 2d^3$,
etc.). Each species received a cytological formula listing
all of the inversions by which it differed from the *D.
repleta* standard. The salivary gland chromosomes, therefore,
contain within themselves a complete record of all of the
successful cytological mutations which have survived during

the history of the species.   This living fossil record allows
us to determine phylogenetic relationships among the species.

II   Results

Figure 2 shows the relationships among the five subgroups
in the *repleta* species group and the *castanea, canalinea,
dreyfusi* and *aureata* species groups.   The non-*repleta* species

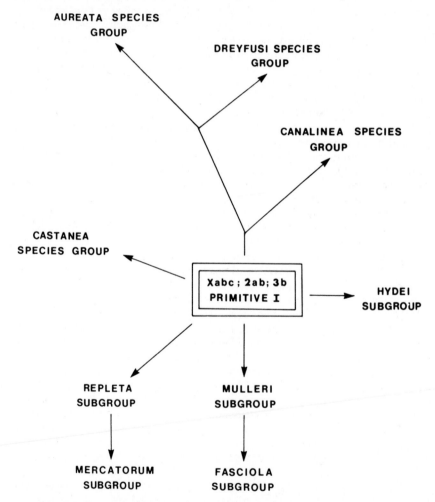

FIGURE 2.   *Phylogenetic relationships within the
Drosophila repleta section.*

groups diverge from the gene sequence which differs from that of *D. repleta* by six inversions, Xa,Xb,Xc,2a,2b, and 3b. This has therefore been chosen as the ancestral type of the *repleta* species group, ANCESTOR I. The phylogeny of the sixty-two species of the *repleta* group are summarized in Wasserman (1982) where specific references are given.

For the most part, relationships are clear-cut and one observes splitting and divergence. The main exception to this is found in the *mulleri* complex, a set of twenty-four species which share, and are homozygous for, eleven inversions. The cytological formula for each of these species is given in Wasserman (1982). The distribution of the shared inversions is given in Table II. It is impossible to explain these data by simple geographical divergence. For example, *D. mojavensis* is homozygous for inversion 2h which is found elsewhere only in *D. arizonensis*, and it is also homozygous for inversion 3a which is lacking in *arizonensis*, but found everywhere else in the *mulleri* complex except in the *buzzatii* cluster (Table II). Similarly among the South American forms (populations Fm, Fs, and Fb in Table II), *martensis* (Fm), is homozygous for Xj and 3a which are only found among the other *martensis* cluster species (Fs), and it is also homozygous for $2e^2$ which is otherwise only known from the *buzzatii* cluster species (Fb). To explain this sharing of inversions we have suggested that the ancestor of the *mulleri* complex, ANCESTOR II, was widely distributed throughout the New World. It consisted of semi-isolated populations between which some gene exchange in the form of migrants took place, but which were sufficiently isolated so that a certain amount of adaptation to local conditions was possible. An adaptive rearrangement would be incorporated into the gene pool of the population in which it arose, and could spread to neighboring populations if it were also advantageous there. If not, then the inversion would be stopped near the border of the two populations; more distant populations would never get a chance to test this sequence. The result would be a step-like cline of inversions, each inversion occurring in its own specific area, with a certain amount of overlapping of ranges among inversions. Thus, the species would consist of a number of cytologically distinct subspecies.

TABLE II. Distribution of Shared Inversions Among
mulleri Complex Species

Ancestral cytological subspecies	Current species	X		Chromosome 2							Chromosome 3	
		j	w	c	f	g	h	d^2	s^6	e^2	a	c
D	arizonensis			+	+	+	+					
C	mojavensis			+	+	+	+				+	
H	from Navojoa			+	+	+					+	
B	mulleri			+	+	+					+	+
	aldrichi			+	+	+					+	+
	wheeleri			+	+	+					+	+
	from Venezuela			+	+	+					+	+
E	spenceri	+	+		+						+	+
	hexastigma	+	+		+						+	+
	longicornis	+	+		+						+	+
	propachuca	+	+		+						+	+
	pachuca	+	+		+						+	+
	mainlandi	+	+		+						+	+
	from Sonora	+	+		+						+	+
A	desertorum	+			+						+	+
	ritae	+			+						+	+
	from Arizona	+			+						+	+
Fs	starmeri	+						+	+		+	
	uniseta	+						+	+		+	
F Fm	martensis	+						+	+	+	+	
Fb	buzzatii							+	+	+		
	serido							+	+	+		
	borborema							+	+	+		

Figure 3 shows the distribution of the shared second chromosome inversions. This chromosome divides the complex into two major parts, the South American forms, populations Fs, Fm and Fb, which have $2d^2s^6$, and the North American forms, the remaining populations, all of which lack $2d^2s^6$ but have 2g. Among the North American forms, populations C and D are the most advanced, having 2g,c,f and h; while the South American forms, Fm and Fb are the most advanced, having $2e^2$ as well as $2d^2s^6$.

Figure 4 gives the distribution of the shared X and third chromosome inversions. All populations have 3a, except the North American population D and the South American

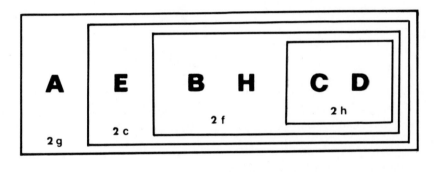

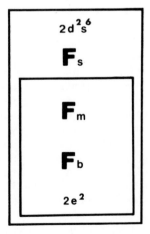

FIGURE 3. *Cytological races of ANCESTOR II as determined by inversions in Chromosome 2.*

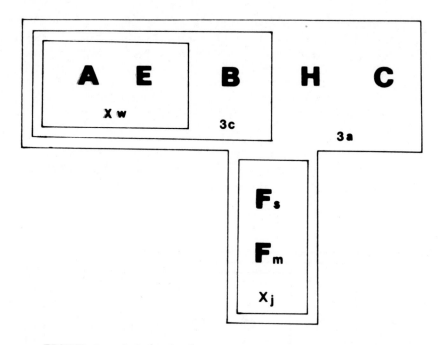

FIGURE 4. *Cytological races of ANCESTOR II as determined by inversions in the X Chromosome and Chromosome 3.*

population Fb. Populations A and E have the most specialized, advanced, X and third chromosomes. Thus we see that the North American population D, which has the most advanced second chromosome, has the most primitive X and third chromosomes. The same is true among the South American forms where population Fb has the most advanced second chromosome and the most primitive X and third chromosomes. When all the data are combined, Figures 5 and 6, a cline of cytologically distinct subspecies ensues.

The *mulleri* complex is now widely distributed throughout the New World but is limited to the cactus deserts where it is a dominant group in the genus. These deserts do not occupy a single continuous area, but are divided into a number of subunits of varying sizes. This is especially true in Mexico, where broken, mosaic patterns of habitats seem to be the rule. Thus the present distribution of the 24 species is probably a reflection and extension of that hypothesized for ANCESTOR II. Figure 5 shows the chromosomal constitution of the Mexican forms. Divergence of the subspecies including further cytological changes, yielded our

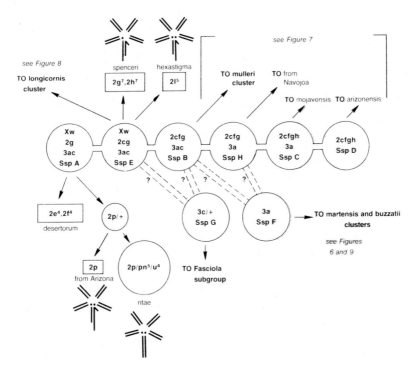

FIGURE 5. *Cytological races of ANCESTOR II from North America.*

modern species. Thus, subspecies D, C and H gave rise to species *arizonensis*, *mojavensis* and "*navojoa*" of the *mojavensis* cluster; while subspecies B gave rise to the *mulleri* cluster (Fig. 7). Subspecies E gave rise to species *hexastigma*, *spenceri* and the four members of the *longicornis* cluster (Fig. 8), while subspecies A yielded the three species of the *ritae* cluster, *ritae*, *tira* and *desertorum*. Subspecies G gave rise to the *fasciola* complex of species. This complex consists of nine species, most of which, unlike the *mulleri* complex, reside in the tropical rainforests of Central and South America (Wasserman, 1962). However, its most primitive member, *D. fulvalineata*, is a desert-inhabiting Mexican species. Finally, the South American section of the *mulleri* complex is derived from subspecies F. Figures 6 and 9 show their cytological evolution. Population Fb evolved into the *buzzatii* cluster with *D. buzzatii* in Brazil, Argentina and Bolivia and *D. serido* and *D. borborema* (not shown) in Brazil. Population Fm evolved into

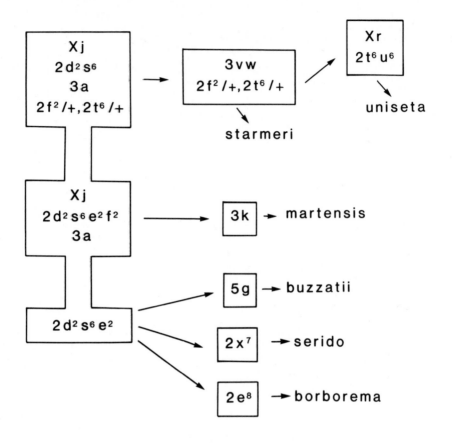

FIGURE 6. Cytological races of ANCESTOR II from South America.

D. *martensis* and Fs into D. *starmeri* and D. *uniseta*, forming the Colombian and Venezuelan species of the *martensis* cluster. Details of the evolution of these forms are given elsewhere: *mojavensis* cluster (Wasserman and Koepfer, 1977a, 1980; Heed, 1978); *longicornis* cluster (Wasserman and Koepfer, 1977b); *martensis* cluster (Wasserman and Koepfer, 1979; Ruiz and Fontdevila, 1982; Ruiz *et al.*, 1982; Fontdevila *et al.*, 1981; Fontdevila, 1982).

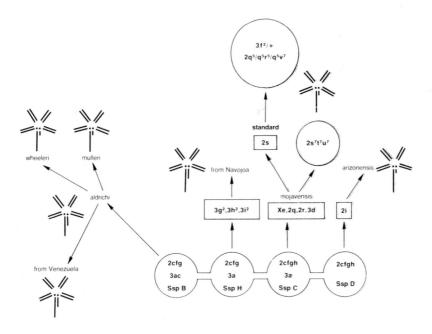

FIGURE 7. *Cytological evolution of the mojavensis and mulleri clusters.*

III Cytological Evolution and Speciation

There has been very little cytological evolution in the *repleta* group. Only a total of 235 inversions has been found to have occurred and survived during the history of 62 species in the *repleta* group (Wasserman, 1982). Of these, 107 are fixed, homozygous differences between species. The average number of fixed, interspecific inversions (1.7) probably will not increase as we accumulate more data unless we are more successful in detecting the smaller overlooked inversions than we have been in the past. In fact, there are six instances involving 17 species, where closely rela- ted species have identical standard chromosomes (Wasserman, 1982). There is an average of about 2.1 polymorphic inver- sions per species (128/62 species). This value will un- doubtedly increase as more data are accumulated. We have found that those species for which three or more localities have been sampled average about six times as much polymor- phic inversions (3.1) as do those species sampled from fewer localities (0.5). Despite the tendency to discover more

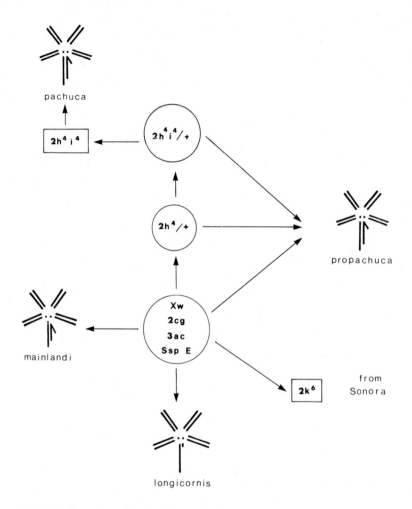

FIGURE 8.   *Cytological evolution of the longicornis cluster.*

polymorphic inversions as new data accumulate, many of the
species are essentially monomorphic and further sampling is
not likely to increase the overall average to much above 3
or so per species.   Moreover, even among those species which
have a considerable amount of inversional polymorphism,
there may be monomorphic populations.   Thus *D. mojavensis*,
which is polymorphic for eight inversions, has a monomorphic
subspecies; *D. nigricruria*, which is polymorphic for four

inversions in Mexico, is monomorphic throughout Central and South America; and *D. mercatorum*, which is polymorphic for nine inversions in Brazil, Bolivia and Chile, is essentially monomorphic elsewhere.

In order to explain the cytological evolution one must take into consideration the following two facts:

(1) Many of the polymorphisms are old. Three inversions are polymorphic in two sister species. Ten other inversions are polymorphic in one species and fixed homozygous in a second species. Thus we see that at least thirteen of the 120 (more than 10%) polymorphic inversions have definitely been polymorphic for a period of time which is longer than that needed for speciation. Others may be equally old, but there are no data concerning their ages.

(2) The inversions are not randomly distributed throughout the genome (Wasserman, 1982). The Inversion Index (number of inversions/unit length of euchromatin) varies considerably: 0.17 (Chromosome 4), 0.27 (Chromosome 5), 0.45 (X), 0.62 (Chromosome 3), to 3.13 (Chromosome 2). Chromosome 2, with 22.6% of the euchromatin, contains 70.2% of the inversions.

What are the mechanisms involved in the evolution of the chromosome? Mutational pressure is obviously not a factor

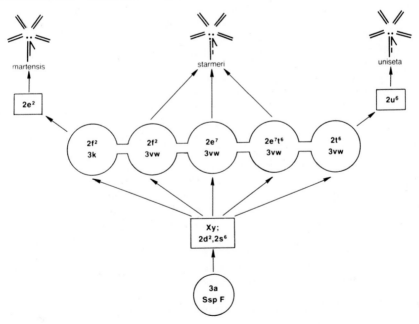

FIGURE 9. *Cytological evolution of the martensis cluster.*

since each inversion is a unique mutation and there is no systematic recurrence of specific inversions. Genetic drift has been implicated as a possible factor establishing geographical differentiation within species and homozygous differences between species (see Hedrick, 1981; Ferrari and Taylor, 1981 for recent discussions). It is generally recognized that drift alone is insufficient and that other factors such as meiotic drive, inbreeding (Hedrick, 1981) or historical events such as glaciations (Krimbas and Loukas, 1980) must work in conjunction with drift. Although drift and meiotic drive may be useful theoretical tools for explaining homozygosity, they cannot reasonably explain the localization of the inversions in Chromosome 2 and their rarity in Chromosomes 4 and 5. Moreover the fact that more than 10% of the polymorphisms are older than the species in which they occur indicates that these inversions are not easily lost even in small populations.

One might suggest that age is a factor. Perhaps speciation involves a bottleneck. One begins with a small isolated population within which rapid changes due to homoselection lead to a subsequent change to a new adaptive peak involving homozygosity for gene orders, as well as the genes themselves. Then, as the species ages, it accumulates new polymorphisms for gene arrangements as well as genes, perhaps through some mechanism similar to the "shifting balance" proposed by Wright (1980). By extension one might also suggest that the number of inversions is indicative of the age of the whole *repleta* species group. Unfortunately the facts belie this suggestion. The degree of polymorphism present in a species does not reflect its age. For example, *D. martensis* is intermediate between *D. starmeri* and *D. buzzatii* (Fig. 6), and therefore is as old or older than, at least, one of the two species. Yet *martensis* is polymorphic for only three inversions while *starmeri* is polymorphic for eighteen and *buzzatii* for twelve (Wasserman, 1982).

The only factor that can operate over such a long period of time in such a systematic way is natural selection. There is ample evidence that inversions are under strong selective pressures including homoselection (Dobzhansky, 1970) Under certain conditions, karyotypic monomorphism must be advantageous and not merely the result of some sampling error that occurs in small populations. Otherwise one must argue that the monomorphic southern race of *D. nigricuria* which occurs in Central and South America is a small population, while the northern polymorphic Mexican race is a large population.

The *repleta* group is quite speciose, yet it has been

cytologically conservative. With an average of only 1.7 homozygous fixed inversional differences between species, the
repleta group has not been subjected to "cytological revolu-
tions" of the sort that White (1978) and Bush *et al.* (1977)
proposed for other organisms. Moreover one finds no correlation between interspecific reproductive isolation and cytological differentiation. *D. mulleri* and *D. aldrichi*, although identical cytologically, mate with a great deal of
reluctance and produce only sterile offspring. On the other
hand, *D. mojavensis* differs from *D. arizonensis* by a minimum
of seven inversions. Yet, under certain laboratory conditions, these species mate readily and produce fertile hybrid
swarms. Zouros (1981) studied the chromosomal basis of the
sexual isolation present between allopatric *mojavensis* and
arizonensis. Although this is lower than that found when
sympatric *mojavensis* is used (Wasserman and Koepfer, 1977a),
Zouros was able to show that discrimination by the males was
controlled by the Y chromosome and either Chromosome 4 or 5;
whereas the discrimination by the females was controlled by
two other autosomes: Chromosomes 2 and either Chromosome 4
or 5 (but the one different from that in the male). *D.
mojavensis* and *D. arizonensis* differ from each other by inversions in the X, 2 and 3, but have cytologically identical
Chromosomes 4 and 5. Yet the latter two chromosomes contain
many of the genes which determine the extent of the reproductive isolation between the two species.

We must conclude, therefore, that although cytological
changes have yielded important information concerning the
evolution of the *repleta* group, they have not been important
in the speciation process.

REFERENCES

Bush, G.L., Case, S.M., Wilson, A.C., and Patton, J.L. (1977).
    Proc. natn. Acad. Sci. U.S.A. 74, 3942.
Dobzhansky, Th. (1970). "Genetics of the Evolutionary
    Process." Columbia University Press, New York and London.
Ferrari, J.A., and Taylor, C.E. (1981). *Evolution 35*, 391.
Fontdevila, A. (1982). Chapter 6, this Volume.
Fontdevila, A., Ruiz, A., Alonso, G., and Ocana, J. (1981).
    Evolution 35, 148.
Hedrick, P.W. (1981). *Evolution 35*, 322.
Heed, W.B. (1978). *In* "Ecological Genetics: The Interface"
    (P.F. Brussard, ed.), p. 109. Springer-Verlag, New York.
Krimbas, C., and Loukas, M. (1980). *Evol. Biol. 12*, 163.

Ruiz, A., and Fontdevila, A. (1982). *Acta cient. venez. 32*, (in press).
Ruiz, A., Fontdevila, A., and Wasserman, M. (1982). *Genetics*, (in press).
Sturtevant, A.H., and Dobzhansky, T. (1936). *Proc. natn. Acad. Sci. U.S.A. 22*, 448.
Ward, C.L. (1949). *Univ. Texas Publ. 4920*, 70.
Wasserman, M. (1962). *Univ. Texas Publ. 6205*, 119.
Wasserman, M. (1963). *Am. Nat. 97*, 333.
Wasserman, M. (1982). *In* "The Genetics and Biology of Drosophila", Volume 3b (M. Ashburner, H.L. Carson and J.N. Thompson, Jr., eds.). (in press).
Wasserman, M., and Koepfer, H.R. (1977a). *Evolution 31*, 812.
Wasserman, M., and Koepfer, H.R. (1977b). *Genetics 87*, 557.
Wasserman, M., and Koepfer, H.R. (1979). *Genetics 93*, 935.
Wasserman, M., and Koepfer, H.R. (1980). *Evolution 34*, 1116.
White, M.J.D. (1978). "Modes of Speciation." W.H. Freeman and Co., San Francisco.
Wright, S. (1980). *Evolution 34*, 825.
Zouros, E. (1981). *Genetics 97*, 703.

5
The Origin of Drosophila in the Sonoran Desert

William B. Heed

Department of Ecology & Evolutionary Biology
University of Arizona
Tucson, Arizona

I Introduction

The origin of the *Drosophila* species of the Sonoran
Desert is of interest for a number of reasons, chief among
them being the question concerning the uniqueness of the
desert as a habitat for these small dipterans. The cactophilic breeding habit in the genus *Drosophila* evolved at
least twice in the Western Hemisphere (Ward and Heed, 1970),
probably in areas considered only as semi-deserts or even
thorn forests today and it therefore represents at most the
major pre-adaptation to desert conditions. Yet the fact that
there are only four species of *Drosophila* unique to the
Sonoran Desert shows the presence of considerable contraints
to diversification (Heed, 1978). Another illustration of
constraints is the realization that the four species evolved
with the desert independently in the sense that the closest
relatives of each one have their major populations either
entirely outside the desert or there is some overlap with the
desert (Heed, 1978). Therefore each species presumably
evolved its own series of adaptations to desert conditions.
The limited number of desert forms permits in-depth comparisons among themselves and among their respective relatives
for a variety of disciplines. The one discussed here is
biogeography.

The present analysis is based on the extensive works of
J. T. Patterson and his students on the Drosophilidae of
southwestern United States and Mexico (Patterson and Crow,
1940; Patterson and Wheeler, 1942; Wharton, 1942, 1943;

ECOLOGICAL GENETICS AND EVOLUTION
ISBN 0 12 078820 9

Patterson and Wagner, 1943; Patterson and Mainland, 1944;
Wheeler, 1949). These works include species descriptions,
classifications, distributions and karyotypes of many
Drosophila of the area. This vast amount of information is
reviewed and up-dated in Patterson and Stone (1952). Since
then the cytological evolution of the gene arrangements in
the polytene chromosomes of many of the species, and newly
described ones, has been worked out for the large cactophilic
repleta species group by Wasserman (1954, 1962, 1982) and for
the small cactophilic nannoptera species group by Ward and
Heed (1970). The uniqueness of paracentric inversions for
determining species relationships in the order Diptera has
been emphasized by Stone (1962) and more recently by Spieth
and Heed (1972).

The species that are endemic to the Sonoran Desert are
listed below according to the species group and complex to
which they belong (Wasserman, 1962, 1982; Ward and Heed,
1970; Heed, 1977). The separate categories underscore their
independent evolutionary pathways.

D. pachea	Patterson & Wheeler	nannoptera species group
D. nigrospiracula	Patterson & Wheeler	repleta species group, anceps complex
D. mettleri	Heed	repleta species group, eremophila complex
D. mojavensis	Patterson & Crow	repleta species group, mulleri complex

The distributional limits of the four species within the
Sonoran Desert as seen in Figures 1-4 are based when necessary on the limits of their host plants as pictured in
Hastings *et al.* (1972). According to Gibson and Horak (1978)
the columnar cacti in the Sonoran Desert represent the end
products of their separate evolutionary paths with relatives
concentrated either in the Nueva Galicia area of southeastern
Mexico for the Stenocereinae or the Puebla-Oaxaco highlands
for the Pachycereinae. The cacti and several of the
Drosophila show approximately the same pattern. Furthermore,
there are generalist and specialist species of yeasts
associated with the columnar cacti (Holzschu and Phaff,
1982). The former occur in southern Mexico while both occur
in the Sonoran Desert.

II D. pachea

The relationship of the species in the nannoptera species group shown in Figure 1 is based chiefly on male and female internal reproductive characters as determined by Lynn Throckmorton (unpublished). According to Throckmorton, the nannoptera species group shows the usual pattern of mosaic evolution with *D. nannoptera* exhibiting a great pre-ponderance of primitive states while *D. pachea* and

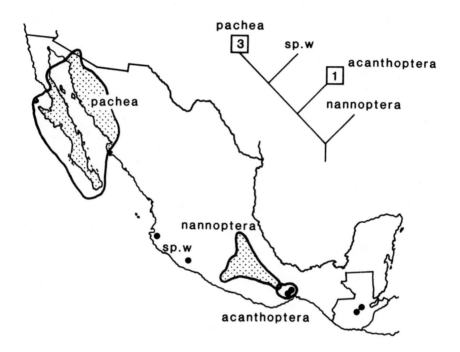

FIGURE 1. Distribution of the nannoptera species group. Localities in Duncan (1979), Heed and Kircher (1965), Heed et al. (1968), Rockwood-Sluss et al. (1973), Russell et al. (1977), Ward and Heed (1970), Ward et al. (1975) and Wheeler (1949). Additional localities and collectors for D. nannoptera: Zumpango, Gro. (M.W.); Mexcala and Chilpancingo, Gro. (R.L.M.). D. acanthoptera: Totolapan, Oax. (R.L.M.); Ixtepec and Tehuantepec, Oax. (J.S.R., R.L.M.). D. "species W": Ixtepec and Tehuantepec, Oax. (J.S.R., R.L.M.); Santa Cruz and Zacapa, Guat. (R.L.M.); Tomatlán, Jal. and Capito, Mich. (W.B.H., E.R.H.).

D. "species W" are more derived and anatomically very close.
D. "species W" is an undescribed form originally recognized
from the Isthmus of Tehuantepec and collected by
M. Wasserman. This scheme was selected in place of the
original relationships as illustrated in Ward and Heed (1970)
because it is also concordant with the chromosome information. In Throckmorton's interpretation, the metaphase of
D. acanthoptera is considered ancestral and the addition and
loss of heterochromatin on various chromosomes produces the
appropriate metaphase for the other species. The numbers in
the boxes in Figure 1 represent inversions that are fixed in
each species. Notice that *D. nannoptera* and *D.* "species W"
are homosequential.

The ecology is very interesting in relation to the
morphological data. The ancestral species, *D. nannoptera*,
has been reared from a variety of arborescent cacti found
in the following genera: *Myrtillocactus*, *Escontria*,
Pachycereus and *Stenocereus* as defined by Gibson and Horak
(1978) while the most derived species, *D. pachea*, is
nutritionally restricted to the genus *Lophocereus*. The other
two species are intermediate in their choice of host plants
in that they utilize chiefly various species in the genus
Stenocereus. This is the only cactophilic species group
that is confined entirely to columnar cacti.

D. nannoptera has a general distribution in the highlands of southern Mexico while *D.* "species W" is in the lowlands and has a more widespread distribution but it is
apparently only locally abundant since the species is not
commonly collected. *D. acanthoptera* is restricted to
southern Oaxaca in the region of the Isthmus of Tehuantepec
where it overlaps *D. nannoptera* at the edge of the Oaxacan
Upland and *D.* "species W" at the Isthmus. *D. pachea* is
disjunct from *D.* "species W" by more than 550 kilometers
from Tomatlán, Jalisco, to Cabo San Lucas in Baja California.
If Baja California were returned to its place of origin at
about 4-6 million years ago (Gastil *et al.*, 1975), then the
disjunction would almost disappear. Alternatively, *D.*
"species W", or its stem population, could have been further
north in the past. In any event the connection to the
Sonoran Desert for this group appears to be along the Pacific
Coast rather than through the highlands of western Mexico.

Therefore the distribution of the nannoptera group
species shows a progression from the ancestral type to the
more derived species following a path from the Neovolcanic
Plateau and the Sierre Madre del Sur (*D. nannoptera*) toward
the lowlands in the region of the Isthmus (*D. acanthoptera*)
and thence east to Guatemala and west to Jalisco

(*D.* "species W"). *D. pachea* was probably derived from a stem population somewhere along the west coast between coastal Jalisco and northern Sinaloa.

Host plant specificity became increasingly specialized through time.

III *D. nigrospiracula*

The inversion phylogeny of the anceps complex is shown in Figure 2 and is based on Wasserman's analysis (1982). The metaphases are summarized in Patterson and Stone (1952). All three species in the anceps complex have three inversions in common while *D. nigrospiracula* became fixed for six unique inversions. No heterozygous inversions have been detected in this species (Cooper, 1964). *D. anceps* has three unique gene sequences while *D. leonis* has one. In addition to these, inversion 2w is fixed in *D. anceps* and is

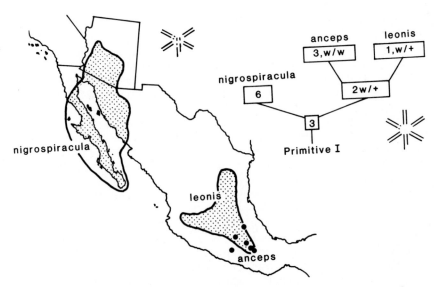

FIGURE 2. Distribution of the anceps complex. Localities in Cooper (1964), Heed (1978), Heed et al. (1968), Jefferson et al. (1974), Patterson and Mainland (1944), Sluss (1975) and Wasserman (1962, 1967). Additional localities and collectors for D. anceps: Zumpango, Gro. (M.W.); Pachuca, Hgo. (R.H.R.); Huahuapan de León, Oax. (M.W.). D. leonis: Huahuapan de León and Oaxaca, Oax. (M.W.).

heterozygous in *D. leonis*. Other inversions are present
heterozygous in *D. leonis* but they do not aid in determining
relationships. *D. anceps* and *D. leonis* have a common
metaphase characterized by a basal constriction on one of
the autosomes and a dot chromosome with added heterochromatin to make it a rod. *D. nigrospiracula* differs by
having a satellite on the X chromosome and the dots are only
slightly enlarged (Patterson and Stone, 1952; Heed, 1977).
It is probable that *D. nigrospiracula* split off early in the
sequence because the dot chromosome is only slightly differentiated.

The chromosome data is concordant with the distribution
of the three species (Figure 2) since *D. leonis* and
D. anceps are sympatric in southern Mexico while
D. nigrospiracula is isolated in the Sonoran Desert.
D. leonis has been recorded as far north as Hidalgo,
Tamaulipas, and as far west as Guadalajara, Jalisco, and as
far south as Oaxaca, Oaxaca. It is sympatric with *D. anceps*
in the Oaxacan Upland. *D. anceps* has also been taken in
Zumpango, Guerrero, where M. Wasserman collected it in
association with *Pachycereus weberi* (pers. comm.). This is
the only host plant record for *D. anceps* and there are no
records for *D. leonis*.

The host plants for *D. nigrospiracula* are *Carnegiea
gigantea* in Arizona and Sonora, *Pachycereus pringlei* in
coastal Sonora and Baja California and *P. pecten-aboriginum*
in the cape region of Baja California (Fellows and Heed,
1972).

The geographic distribution of *P. pecten-aboriginum*,
aside from the cape region, extends along the foothills of
the Pacific coast from southern Sonora to the Isthmus of
Tehuantepec in Oaxaca (Bravo-Hollis, 1978) and therefore the
cactus may be considered a prime candidate for moving the
anceps complex into, or away from, the Sonoran Desert.
However, there are no records of *D. nigrospiracula*, or a
close associate, outside of the Sonoran Desert along the
Pacific coast. Furthermore, *D. nigrospiracula* has never
been reared from this cactus on the mainland.

The anceps complex has two points in common with the
nannoptera species group. They both exhibit disjunctions
from Sonora and Baja California to southern Mexico and their
species have only been reared from columnar cacti. The
anceps complex differs from the nannoptera group in that
the southern relatives are restricted to the highlands. A
further difference is the apparent early split in the anceps
complex that eventually led to *D. nigrospiracula* and a late
split in the nannoptera group to form *D. pachea*.

IV *D. mettleri*

The eremophila complex was formulated by Wasserman (1982) and it contains only two species (Figure 3). *D. mettleri* and *D. eremophila* have four inversions in common while the latter species is fixed for two inversions unique to it. The metaphase of *D. eremophila* from Acatlán, Puebla, has a small pair of V's in place of the dot chromosomes and a very short Y chromosome (Wasserman, 1962).

The Y chromosome of *D. eremophila* from Guayalejo, Tamaulipas, is longer than the Acatlán Y but it is less than

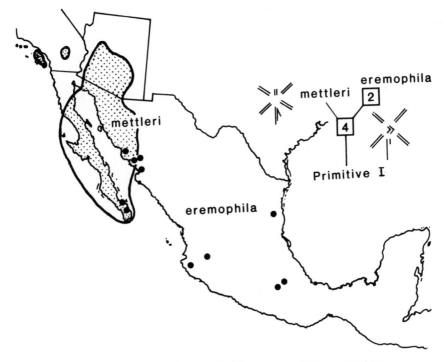

FIGURE 3. Distribution of the eremophila complex. Localities in Fogleman et al. (1982), Heed (1977, 1978), Jefferson et al. (1974), Wasserman (1962, 1967). Additional localities and collectors for D. eremophila: Tomatlán and Guadalajara, Jal. (W.B.H., R.H.T.); La Paz and San Bartolo, B.C.S. (W.B.H., W.T.S.); Los Mochis, Sin. (W.B.H., J.S.J.); Empalme, Son. (W.B.H., J.S.J.); Alamos, Son. (A.R., S.J.). D. mettleri: Santa Catalina Island, Calif. (W.B.H., M.E.H.); Vallecito, Calif. (W.B.H., J.C.F.).

half the length of the X, while it is over half the length of
the X in a strain from Navojoa, Sonora (Heed, 1977). The
metaphase of *D. mettleri* from Tucson, Arizona, has a pair of
short rods in place of the dot chromosome and a large
J-shaped Y chromosome (Heed, 1977).

   D. eremophila has a wide distribution in Mexico and has
been taken in the lowlands, as well as the highlands, at
Guadalajara, in Jalisco. *D. eremophila* is sympatric with
D. mettleri in the cape region of Baja California and in
southern Sonora. The species is associated with prickly pear
cactus but has not been reared from any definite substrate
probably because it has the same soil breeding habit as
D. mettleri. This species follows the same host plants as
D. nigrospiracula in the Sonoran Desert but it breeds in the
moist soil beneath the rotting arms of these giant cacti
(Heed, 1977). It has also been bred from the rot-soaked soil
of *Stenocereus thurberi* in the desert (Fogleman *et al.*, 1981)
and adults are often aspirated along with *D. pachea* from the
rotting tissue of *Lophocereus schottii* (Fogleman *et al.*,
1982). In the Anza-Borrego Desert of southern California,
several specimens of *D. mettleri* were collected in association with the California Barrel Cactus (*Ferocactus
acanthodes*), in the winter of 1979 and again in the spring
of 1981. The species was very abundant on Santa Catalina
Island, California, in early November, 1981, where it was
associated with *Opuntia "demissa"* and *O. oricola*. Many flies
were reared from the fermenting fruits and pads but none of
them were *D. mettleri*. Therefore it is believed the species
is a soil breeder under the Opuntias also.

   It is difficult to interpret the history of these two
species because they both have derived metaphase chromosomes
and not many inversions. However, the polytypism in the
length of the Y chromosome in *D. eremophila* could be
significant because it is most similar to *D. mettleri* where
the species are sympatric and least similar in the area of
Acatlán, Puebla. This suggests the common ancestor existed
in the past in northwestern Mexico. The probable soil
breeding habit of *D. eremophila* would permit invasion southward without competition from other species.

V   *D. mojavensis*

   D. mojavensis is a member of the large mulleri complex of
species that extends to South America. The inversion phylogeny in Figure 4 shows only its closest relatives as analyzed

by Wasserman (1962, 1982). Inversion 2h is critical for it splits out *D. mojavensis* and *D. arizonensis* from the remainder of the mulleri complex, including *D.* "species N", an undescribed form originally recognized from Navojoa, Sonora. *D. mojavensis* then became fixed for five inversions,

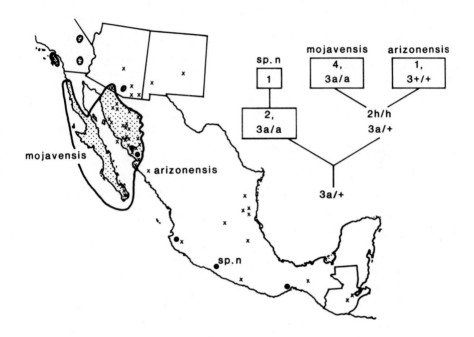

FIGURE 4. *Distribution within the mulleri complex. Localities in Fellows and Heed (1972), Heed (1978), Heed et al. (1968), Jefferson et al. (1974), Johnson (1980), Markow (1981), Mettler (1957, 1963), Patterson and Mainland (1944), Richardson et al. (1977), Spencer (1941), Starmer et al. (1977), Wasserman (1962), Wasserman et al. (1971), Zouros (1973), Zouros and Johnson (1976), and Zouros and d'Entremont (1980). Additional localities and collectors for D. arizonensis: Santa Cruz and Zacapa, Guat. (R.L.M.); Tuxtla Gutiérrez, Chps. (M.W.); Tomatlán, Jal. and Guamúchil, Sin. (W.B.H., R.H.T.); Los Planes, B.C.S. (W.B.H., A.R.); La Paz and San Bartolo, B.C.S. (W.B.H., W.T.S.). D. mojavensis: Vallecito, Calif. (W.B.H., J.C.F.); Santa Catalina Island, Calif. (W.B.H., M.E.H.); Nátora, Son. (J.S.R., A.R.). D. "species N": Tehuantepec, Oax. (M.W.); Zijhuatanejo, Gro. (M.W.); Tomatlán, Jal. (W.B.H., R.H.T.); Navojoa, Son. and Los Mochis, Sin. (W.B.H., J.S.J.).*

including 3a, while *D. arizonensis* lost 3a and fixed only one
other inversion. *D.* "species N" has one more inversion
homozygous in Navojoa, Sonora, than in the three localities
along the coast of southern Mexico (Figure 4) where there are
three fixed sequences, including 3a (Wasserman, pers. comm.).
D. mojavensis is polymorphic for a number of inversions that
are restricted mostly to Baja California (Mettler, 1963;
Johnson, 1980). *D. arizonensis* has no known inversion polymorphism. The metaphase chromosome of all three species are
generally similar. They have the usual five pairs of rods
and a pair of dots. In Navojoa, Sonora, *D. mojavensis* and
D. "species N" have a Y chromosome that is one-half the
length of the X. The length of the Y chromosome in
D. arizonensis from the Rio Cuchujaqui in Sonora is at least
three quarters as long as the X (Heed, unpublished). These
determinations for the length of the Y differ from the
original descriptions in Wharton (1942, 1943) where
D. mojavensis (from Death Valley, California) and
D. arizonensis (from Arizona) are pictured as having a very
short Y.

The distributions of the three species can be seen in
Figure 4. *D. arizonensis* has the widest distribution from
Phoenix, Arizona, and Ruidosa, New Mexico, to the Rio Motagua
Basin in Guatemala. It has also been taken in low numbers on
four separate occasions in the cape region of Baja California. The species apparently occurs in both the lowlands and
highlands. *D. mojavensis* occurs throughout the Sonoran
Desert except in the major part of southern Arizona and in
extreme northern Baja California. According to
Warren Spencer (1941) the species may be commonly encountered
in the Mojave Desert. *D. mojavensis* has been taken also in
various localities in southern California by M. Wasserman in
the summer of 1975 (pers. comm.). Recently, collectors from
the University of Arizona have discovered *D. mojavensis* in
the Anza-Borrego Desert and Santa Catalina Island in California. *D.* "species N" is a lowland coastal form that has been
collected sporadically from Navojoa, Sonora, to the Isthmus
of Tehuantepec.

D. mojavensis breeds in a variety of cacti in the Sonoran
Desert but its chief host plants are *Stenocereus gummosus* and
Stenocereus thurberi (Fellows and Heed, 1972; Heed, 1978).
The species was first associated with *Ferocactus acanthodes*
in the Mojave Desert (Spencer, 1941). *D. mojavensis* has been
reared since from this species of barrel cactus from the
Anza-Borrego Desert (Heed, unpublished). On Santa Catalina
Island, the species was reared mostly from the fruits of
Opuntia "*demissa*" and to a lesser extent from the rotting

pads. This is the first substantial record of *D. mojavensis* breeding in *Opuntia*. *D. arizonensis* breeds mostly in *Stenocereus alamosensis* in Sonora (Fellows and Heed, 1972) and different species of *Opuntia* in southern Sonora, Arizona, and probably New Mexico. It has been reared from *Opuntia wilcoxii* in Guamúchil, Sinaloa, and from a columnar cactus tentatively identified as *Stenocereus fricii* in Tomatlán, Jalisco (Heed, unpublished). In Tamaulipas, San Luis, Potosi and Hidalgo, Richardson *et al.*, (1977) associate *D. arizonensis* with *Myrtillocactus geometrizans*. *D.* "species N" breeds in *Opuntia wilcoxii* in southern Sonora and in Tomatlán, Jalisco (Heed, unpublished). However, in Tomatlán the species prefers the fruits instead of the rotting pads. There are no *Opuntia* fruit records from southern Sonora for any species.

The evolution of the mulleri complex in tropical America has been discussed by Wasserman on several occasions (Wasserman, 1960, 1962, 1982) and the geographical speciation of *D. arizonensis* and *D. mojavensis* in particular has been recently explored with a variety of models (see Wasserman and Koepfer, 1980, for a review). The basic data on the degree of behavioral isolation and of the genetic difference between the two species (Baker, 1947; Crow, 1942; Markow, 1981; Mettler, 1957; Nagle and Mettler, 1969; Richardson *et al.*, 1977; Wasserman and Koepfer, 1977; and Zouros, 1973, 1981a, 1981b) and between two of the races of *D. mojavensis* (Heed, 1978; Zouros, 1973; Zouros and d'Entremont, 1980) are mentioned here to emphasize that the process of species splitting, if captured at the proper moment in evolutionary time, can be analyzed with great precision. The question of the origination of *D. mojavensis*, however, has been investigated most thoroughly by Johnson (1980) who analyzed its inversion polymorphism and behavior in the field. Johnson considers the species to have originated in Baja California for the following reasons: (1) all seven polymorphic inversions probably arose in the peninsula, (2) the vast majority of the polymorphic populations reside there, (3) a relictual homologue, SI, having three inversions, has been uncovered there on two separate occasions, (4) the preferred host plant (*Stenocereus gummosus*) is most abundant in the peninsula and (5) all secondary host plants also can be found in Baja California.

The three species within the mulleri complex exist sympatrically in southern Sonora and northern Sinaloa (Figure 4). This is believed to be an area of secondary overlap partly because *D.* "species N" and *D. mojavensis* are both cytologically derived in the area. The origin of

D. arizonensis is more difficult to interpret because of its uniform cytological characteristics. Consideration of its very close genetic relatedness to *D. mojavensis*, however, must place the origin of the species on the mainland near where the Sonoran Desert exists today. It spread north and south from that area. The ancestral type probably invaded *Stenocereus alamosensis* from an *Opuntia* type in Sonora or Sinaloa. Alternatively *S. kerberi* in Sinaloa was probably also available. Both cactus species extend southward at least to the state of Colima in a very probably patchy distribution. This possibility, and the fact that it is adapted to more humid conditions than is *D. mojavensis* (Fellows and Heed, 1972), may have permitted *D. arizonensis* to disperse south where it exists in other parts of Mexico in *Myrtillocactus* and other species of *Stenocereus*.

VI   Discussion and Conclusions

The analysis of the origination of the species of *Drosophila* endemic to the Sonoran Desert is difficult primarily because the desert itself is rather refractive to historical investigation. There is little paleobotanical information on the arid regions of Mexico (Rzedowski, 1973; Van Devender, 1977; Axelrod, 1979). In addition, more study is needed on the distributions and relationships of the columnar cacti, not only in Mexico, but also especially from South America (Gibson and Horak, 1978). For instance, *Stenocereus chrysocarpus* and *S. fricii* were not described until 1972 and 1973, respectively (Bravo-Hollis, 1978). The mulleri complex is undergoing intensive investigation in South America at the present time (Sene *et al.*, 1982; Fontdevila, 1982). In northwestern Mexico there are critical areas that remain uncollected for *Drosophila* such as the narrow coastal strip in Sinaloa and the mountains and foothills in Sonora, Sinaloa, Chihuahua and Durango. It would also be helpful to have the internal reproductive system of each species analyzed in the manner described by Throckmorton (1962) and as he examined the nannoptera group here. In addition, events in the evolution of the family Drosophilidae (Throckmorton, 1975, 1982) need to be connected more strongly with the events occurring at the more local level described here. Also, the breeding ecology of the southern species is known only superficially. The *Drosophila* of the Sonoran Desert show strong host plant preferences and they are reflected in the

distribution of the species of yeasts among the columnar
cacti (Starmer *et al.*, 1980). Studies of this same magnitude
in southern Mexico should prove useful in determining the
relative strength of host preference in that area. The
relation between the yeast species of *Opuntia* and the
columnar cacti are reviewed by Starmer (1981). Finally the
evolutionary history of *D. mojavensis* is much more extensive
than outlined in the present paper. For instance there are
four other species in the mulleri complex found at the edges
of the Sonoran Desert and they tell a different story.

It is most probable that in more pluvial times the
Opuntia breeders, *D. aldrichi* and *D. longicornis*, or their
precursors, had more extensive distributions that reached
across the Sonoran Desert to the coastal sage association of
the Pacific coast in the region of southern California and
northern Baja California. The species that represent them
today in that locality, and still breeding in *Opuntia*, are
D. wheeleri and *D. mainlandi* respectively, having been
isolated in that general region as conditions became drier in
the interior. Today *D. aldrichi* is found no further north
than the thorn scrub of southern Sonora and *D. longicornis*
apparently does not extend to the Arizona-California border,
although it is present around Tucson. *D. wheeleri* and
D. mainlandi also exist sympatrically in the cape region of
Baja California. These events were probably not simultaneous
with the origin of *D. mojavensis* and *D. arizonensis*, as
described previously. They suggest the mulleri complex
evolved over a period that exhibited considerable climatic
change in the southwestern U.S. and northwestern Mexico. For
a review of the probable effects of alternating moist and dry
periods in the tropics during the Pleistocene, and before,
see Axelrod (1979), Raven and Axelrod (1974) and Webb (1978).
For a concise history of the evolution of the saguaro, see
Lowe and Steenbergh (1981).

Regardless of the uncertainties enumerated above, the
four species of *Drosophila* restricted to the desert can be
combined in various ways in an attempt to determine their
relative ages, among other things. The nannoptera species
group and the anceps complex have the columnar breeding
habit as far as the records indicate and they have the large
disjunctions from the Sonoran Desert to the state of Jalisco.
Thus the two species may be of greater age than either
D. mettleri or *D. mojavensis*. Long distance dispersal from
the southern to the northwestern part of Mexico of course is
possible but little would be learned concerning the evolution
of arid lands if this explanation were invoked whenever
convenient. Also *D. pachea* and *D. nigrospiracula* do not

extend into the deserts of southern California presumably because they could not make the shift from the columnars to the prickly pears and the barrels.

The eremophila complex and the three species at the tip of one branch of the mulleri complex both show overlap of species distributions chiefly in the thorn scrub of southern Sonora. Prickly pear cactus is a prominent host plant in both groups. Also *D. mojavensis* and *D. mettleri* have extended their ranges into southern California and probably very recently to Santa Catalina Island. These may be features of younger polyphagic species. However, in regard to phylogenetic trends, *D. pachea* and *D. mojavensis* are obviously derived species and they show evidence of origin from the south along the coast while *D. nigrospiracula* and *D. mettleri* are neither clearly derived forms, compared to their relatives, nor is the route to their place of origin at all certain. From these comparisons, *D. nigrospiracula* comes out as an older species on both counts while *D. mojavensis* emerges twice as a younger one. This is at least a beginning in our understanding of the origins of these flies.

ACKNOWLEDGMENTS

The collectors listed in the figures are: James C. Fogleman, Emily R. Heed, M. Ellen Heed, J. Spencer Johnston, Steven Jones, Robert L. Mangan, Richard H. Richardson, Alexander Russell, Jean S. Russell, William T. Starmer, Richard H. Thomas, Marvin Wasserman and the author. Several cactus specimens were kindly identified by Arthur C. Gibson and Donald J. Pinkava.

The paper is dedicated in the memory of Alexander ("Ike") Russell who generously gave of his time and savvy of the Sonoran Desert. The work was supported by NSF.

REFERENCES

Axelrod, D.I. (1979). *Occ. Papers Calif. Acad. Sci. 132*, 1.
Baker, W.K. (1947). *Univ. Texas Publ. 4720*, 126.
Bravo-Hollis, H. (1978). "Las cactáceas de México." Univ.
      Nac. Autón. de Mex., Mexico.
Cooper, J.W. (1964). M.S. Thesis, Univ. of Arizona.
Crow, J.F. (1942). *Univ. Texas Publ. 4228*, 53.
Duncan, G.A. (1979). Ph.D. Dissertation, Univ. of Arizona.

Fellows, D.P., and Heed, W.B. (1972). *Ecology 53*, 850.

Fogleman, J.C., Hackbarth, K.R., and Heed, W.B. (1981). *Am. Nat. 118*, 541.

Fogleman, J.C., Heed, W.B., and Kircher, H.W. (1982). *Comp. Biochem. Physiol.* (in press).

Fontdevila, A. (1982). Chapter 6, this Volume.

Gastil, R.G., Phillips, R.P., and Allison, E.C. (1975). "Geol. Soc. of Amer. Memoir 140." Geol. Soc. Amer. Inc., Boulder, Colorado.

Gibson, A.C., and Horak, K.E. (1978). *Ann. Missouri Bot. Gard. 65*, 999.

Hastings, J.R., Turner, R.M., and Warren, D.K. (1972). "Tech. Rep. on the Meteor. and Climat. of Arid Regions, No. 21." Inst. Atm. Phys., Univ. of Arizona, Tucson.

Heed, W.B. (1977). *Proc. ent. Soc. Wash. 79*, 649.

Heed, W.B. (1978). *In* "Ecological Genetics: The Interface" (P.F. Brussard, ed.) p. 109. Springer-Verlag, New York.

Heed, W.B., and Kircher, H.W. (1965). *Science 149*, 758.

Heed, W.B., Russell, J.S., and Ward, B.L. (1968). *Drosophila Inf. Serv. 43*, 94.

Holzschu, D.L., and Phaff, H.J. (1982). Chapter 9, this Volume.

Jefferson, M.C., Johnson, W.R. Jr., Baldwin, D.G., and Heed, W.B. (1974). *Drosophila Inf. Serv. 51*, 65

Johnson, W.R. Jr. (1980). Ph.D. Dissertation, Univ. of Arizona.

Lowe, C.H., and Steenbergh, W.F. (1981). *Desert Plants 3*, 83.

Markow, T.A. (1981). *Evolution 35*, 1022.

Mettler, L.E. (1957). *Univ. Texas Publ. 5721*, 157.

Mettler, L.E. (1963). *Drosophila Inf. Serv. 38*, 57.

Nagle, J.J., and Mettler, L.E. (1969). *Evolution 23*, 519.

Patterson, J.T., and Crow, J.F. (1940). *Univ. Texas Publ. 4032*, 251.

Patterson, J.T., and Mainland, G.B. (1944). *Univ. Texas Publ. 4445*, 9.

Patterson, J.T., and Stone, W.S. (1952). "Evolution in the Genus Drosophila." MacMillan Co., New York.

Patterson, J.T., and Wagner, R.P. (1943). *Univ. Texas Publ. 4313,* 217.

Patterson, J.T., and Wheeler, M.R. (1942). *Univ. Texas Publ. 4213*, 67.

Raven, P., and Axelrod, D.I. (1974). *Ann. Missouri Bot. Gard. 61*, 539.

Richardson, R.H., Smouse, P.E., and Richardson, M.R. (1977). *Genetics 85*, 141.

Rockwood-Sluss, E.S., Johnston, J.S., and Heed, W.B. (1973). *Genetics 73*, 135.

Russell, J.S., Ward, B.L., and Heed, W.B. (1977). *Drosophila Inf. Serv. 52*, 112.

Rzedowski, J. (1973). *In* "Vegetation and Vegetational History of Northern Latin America" (A. Graham, ed.), p. 61. Elsevier Sci. Publ. Co., Amsterdam.

Sene, F.de M., Pereira, M.A.Q.R., and Vilela, C.R. (1982). Chapter 7, this Volume.

Sluss, E.S. (1975). Ph.D. Dissertation, Univ. of Arizona.

Spencer, W.P. (1941). *Ohio J. Sci. 41*, 190.

Spieth, H.T., and Heed, W.B. (1972). *Annu. Rev. Ecol. & Syst. 3*, 269.

Starmer, W.T. (1981). *Evolution 35*, 38.

Starmer, W.T., Heed, W.B., and Rockwood-Sluss, E.S. (1977). *Proc. natn. Acad. Sci. U.S.A. 74*, 387.

Starmer, W.T., Kircher, H.W., and Phaff, H.J. (1980). *Evolution 34*, 137.

Stone, W.S. (1962). *Univ. Texas Publ. 6205*, 507.

Throckmorton, L.H. (1962). *Univ. Texas Publ. 6205*, 207.

Throckmorton, L.H. (1975). *In* "Handbook of Genetics" (R.C. King, ed.), Vol. 3, p. 421. Plenum, New York.

Throckmorton, L.H. (1982). Chapter 3, this Volume.

Van Devender, T.R. (1977). *Science 198*, 189.

Ward, B.L., and Heed, W.B. (1970). *J. Hered. 61*, 248.

Ward, B.L., Starmer, W.T., Russell, J.S., and Heed, W.B. (1975). *Evolution 28*, 565.

Wasserman, M. (1954). *Univ. Texas Publ. 5422*, 130.

Wasserman, M. (1960). *Proc. natn. Acad. Sci. U.S.A. 46*, 842.

Wasserman, M. (1962). *Univ. Texas Publ. 6205*, 85.

Wasserman, M. (1967). *Drosophila Inf. Serv. 42*, 67.

Wasserman, M. (1982). *In* "The Genetics and Biology of Drosophila" Vol. 3b. (M. Ashburner, H.L. Carson and J.N. Thompson, Jr., eds.), Academic Press, New York (in press).

Wasserman, M., and Koepfer, H.R. (1977). *Evolution 31*, 812.

Wasserman, M., and Koepfer, H.R. (1980). *Evolution 34*, 116.

Wasserman, M., Koepfer, H.R., and Geller, M.J. (1971). *Drosophila Inf. Serv. 46*, 122.

Webb, D.S. (1978). *Annu. Rev. Ecol. & Syst. 9*, 393.

Wharton, L.T. (1942). *Univ. Texas Publ. 4228*, 23.

Wharton, L.T. (1943). *Univ. Texas Publ. 4313*, 282.

Wheeler, M.R. (1949). *Univ. Texas Publ. 4920*, 157.

Zouros, E. (1973). *Evolution 27*, 601.

Zouros, E. (1981a). *Genetics 97*, 703.

Zouros, E. (1981b). *Can. J. Genet. Cytol. 23*, 65.

Zouros, E., and d'Entremont, C.J. (1980). *Evolution 34*, 421.

Zouros, E., and Johnson, W. (1975). *Can. J. Genet. Cytol. 18*, 245.

6

Recent Developments on the Evolutionary History of the *Drosophila mulleri* Complex in South America

Antonio Fontdevila

Departamento de Genética
Facultad de Ciencias
Universidad Autónoma
Bellaterra (Barcelona)
Spain

I Phylogenetic and Reproductive Relationships

The species of the *mulleri* complex have been combined into clusters in order to better reflect their cytological evolution (Wasserman, 1982a). Wasserman *et al.* (1973) included the South American species *D. martensis*, *D. uniseta* and *D. starmeri* in the *mulleri* complex and Wasserman and Koepfer (1979) established them as the cluster *martensis*. This work on the *martensis* cluster stimulated our interest in studying the evolution of the *mulleri* subgroup in South America.

The precise phylogenetic position of *D. buzzatii*, a South American *mulleri* subgroup species (Patterson and Wheeler, 1942) had always been unclear to me. It was derived by Wasserman (1962) from an ancestral form (PRIMITIVE I) by the fixation of three inversions in the second chromosome and one in the fifth chromosome. Crow (1942) had shown previously that this species was distantly related reproductively to two members of the *mulleri* complex (*D. mulleri* and *D. arizonensis*), because interspecific crosses gave larvae when *D. buzzatii* was the parental male. However, since no cytological affinities were apparent, it was placed, together with *D. pegasa*, a cytologically similar species, in the miscellaneous section of the *mulleri* subgroup. There was no other supporting evidence for a close affinity between *D. buzzatii* and *D. pegasa*.

ECOLOGICAL GENETICS AND EVOLUTION
ISBN 0 12 078820 9

The former originated in South America, whereas the latter is limited to Mexico. The metaphase chromosomes of *D. pegasa* consist of four pairs of autosomal rods, a pair of dumbbell-shaped dot autosomes and a rod-shaped X (the shape of the Y is unknown). On the other hand, the metaphase of *D. buzzatii* differs from that of *D. pegasa* by showing two simple dots and a short rod Y. The species are neither reproductively nor ethologically similar. They do not cross and *D. pegasa* shows a very peculiar courtship behavior.

Recent study of newly collected samples from South America showed that three of the inversions which led to the origin of *D. buzzatii* are identical to three of the inversions found in the *martensis* cluster, i.e., $2d^2$, $2e^2$ and $2s^6$ (Ruiz *et al.*, 1982). Thus, the standard sequence of *D. buzzatii* is Xabc; $2abd^2s^6e^2$; 3b; 5g, differing from the basic *repleta* sequence by 10 inversions. This finding gave us the basis to place this species in the *mulleri* complex, although we needed to redefine it cytologically, since *D. buzzatii* shares none of the inversions originally used to define the *mulleri* complex. Figure 1 outlines the overall cytological evolution of the redefined *mulleri* species. PRIMITIVE I (Xabc, 2ab, 3b) is related to PRIMITIVE II, the

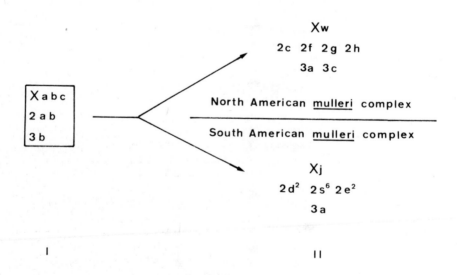

FIGURE 1. Diagram of the cytological evolution and new definition of the mulleri complex in South America. (From Ruiz et al., 1982).

polytypic ancestor of the *mulleri* complex. The North
American species are homozygous for at least one of the
characteristic seven inversions (Xw, 2c, 2f, 2g, 2h, 3a and
3c), but the South American species are homozygous for one
or more of five inversions (Xj, $2d^2$, $2e^2$, $2s^6$ and 3a). The
3a sequence is found in some species of both geographical
regions.

This study showed also an expansion of the range of the
mulleri complex in South America. *D. serido* was found sympatric with *D. buzzatii* in several Argentinian localities
(Ruiz *et al.*, 1982). Cytologically, the primitive standard
second chromosome of these *D. serido* populations can be
derived from the basic rearrangement of *D. buzzatii* simply by
inverting the $2j^9$ segment. The standard fifth chromosome of
D. serido also differs from *D. buzzatii* in that it is the
same as the *repleta* standard (it lacks the 5g inversion).
The rest of the standard chromosomes are the same for both
species. The phylogenetic closeness between these species is
evidenced also by hybridization tests. Interspecific hybrids
are obtained only when the cross is between *D. serido* females
and *D. buzzatii* males. The F_1 progeny do not produce offspring. Hybrid males are sterile, but hybrid females are
fertile and when crossed with males of either parental
species give a few offspring. These findings have given us
support to include *D. serido* and *D. buzzatii* in a new taxonomic unit, the *buzzatii* cluster.

The studies of collections in northern South America also
have been very fruitful. Perusal of the chromosomal polymorphism of *D. starmeri* (Table I) and the map of northern
South America (Fig. 2) reveals that inversions $2z^6$, $2y^6$, $2a^7$,
$2t^6$, $2q^9$, Xs and Xy are restricted to the West of the Andean
Cordillera and the Coastal Ranges of Venezuela (Sierra
Costera) (western region) and also that inversion $2w^6$ is only
found to the east of these mountains (eastern region). Additionally, inversion Xq is very frequent or is fixed in the
western region and is either rare or absent in the eastern
region. This geographical differentiation is sufficient to
define two cytological races of *D. starmeri*. Furthermore,
the reality of these two races is reinforced by intraspecific
hybridization tests (Table II). Whenever the parents belong
to different geographical races, F_2 breakdown is found.
Moreover, in four of these crosses in which the male parent
is from the oriental race, there are no F_2 offspring because
F_1 males are sterile. No F_2 breakdown has been observed in
crosses within a race. These tests demonstrate the advanced
stage of divergence between the two races.

FIGURE 2.   *Geographical localities of several natural*
populations of different species of the mulleri complex
sampled by us in South America: ● *D. buzzatii;* ○ *D. serido;*
▲ *D. starmeri;* ■ *D. martensis;* □ *D. uniseta;* △ *D. species*
"v".   Life zones of Argentina: Ch: Chaco; Pa: Pampa; E.
Espinal; M: Monte; P: Puna

TABLE I. *Geographical Distribution of Chromosomal Arrangements of D. starmeri in Some Populations of Northern South America (from Ruiz and Fontdevila, 1982)(see Figure 2 for Location of Populations)*

Locality	Number of analyzed strains	Chromosomal arrangement				
		Xabcj	$2abd^2s^6$	3abvw	4	5
Eastern region						
1. Guaca	20	St	e^7 $f^2x^6w^6$	St	St	St
4. Zuata	5	St	e^7 $f^2x^6w^6$	St	St	St
Western region						
5. Cata	10	St q	e^7 e^7t^6 $e^7t^6q^9$	St	St	St
6. Lagunillas	12	q	e^7 $f^2x^6z^6$	St	St	St
9. Riohacha	32	q qs	e^7 $f^2x^6z^6$ $f^2x^6z^6a^7$ $f^2x^6y^6z^6$ $f^2x^6z^6a^7y^6$	St	St	St
11. Sta. Rita	1	qy	e^7	St	St	St

These studies revealed also the presence of a new species sympatric with *D. starmeri* in the east of Venezuela (Oriente) and morphologically very similar to this species. Chromosomally, the new species, *D.* species "v" is not distinguishable from *D. starmeri* and the strains analyzed so far reveal that they are homozygous for the sequence: Xabcj, $2abd^2s^6e^7$, 3abvw, 4 and 5. Polytene chromosomal pairing in hybrids is good and asynapsis is observed only in the proximal ends, due probably to different amounts in centromeric heterochromatin. The close relationship between *D.* species "v" and *D. starmeri* also can be seen in interspecific hybridization crosses (Table III). Crosses between *D. starmeri* males and *D.* species "v" females always give abundant F_1 offspring, but the F_1 males are sterile. Reciprocal crosses give some F_1 offspring when the *D. starmeri* parental females belong to the western race, but no offspring if the females are from the

TABLE II.   *Average Number of Progeny (± standard error) in F_2 Crosses among D. starmeri Strains from Several Localities in Northern South America (from Ruiz and Fontdevila, 1982) (see Figure 2 for Location of Populations)*

| | Locality number | | | | |
| | Western region | | | Eastern region | |
	11	9	5	4	1
11	177.0 (±22.7)	102.4 (±18.0)	111.4 (±23.1)	27.8 (±7.6)	77.6 (±11.6)
9	142.8 (± 8.9)	89.0 (± 6.8)	120.2 (±13.4)	–	–
5	74.0 (±25.1)	73.6 (±11.2)	116.6 (±13.3)	–	–
4	30.8 (±13.2)	110.5 (± 8.4)	5.0 (± 3.9)	124.4 (±16.0)	144.2 (± 9.7)
1	15.2 (±13.7)	65.4 (±20.3)	14.8 (± 6.3)	134.7 (± 6.5)	155.4 (±14.7)

TABLE III.   *Average Number of Progeny (± standard error) in F_1 Crosses between D. species "v", and D. starmeri Strains from Several Localities of Northern South America (from Ruiz and Fontdevila, 1982)*

Type of cross (♂♂ x ♀♀)	F_1 progeny
Sta. Rita x D. species "v"	36.2 ± 6.7
D. species "v" x Sta. Rita	9.2 ± 5.6
Riohacha x D. species "v"	24.4 ± 5.3
D. species "v" x Riohacha	2.6 ± 1.5
Cata x D. species "v"	44.4 ± 9.9
D. species "v" x Cata	0.4 ± 0.9
Zuata x D. species "v"	15.2 ± 7.0
D. species "v" x Zuata	–
Guaca x D. species "v"	38.4 ± 13.8
D. species "v" x Guaca	–

eastern race. In either case F_1 males are always sterile, but hybrid females are fertile and give some progeny with males of either parental species.

These new data on the South American section of the *mulleri* complex have unequivocally defined and expanded the *martensis* and the *buzzatii* clusters and also demonstrated their mutual relationships. Figure 3 summarizes the cytological phylogeny of these clusters. Subspecies F appears subdivided in three evolutionary units which share inversions and represent the minimum number of ancestral populations needed to explain the evolution of these clusters. Ancestor IIFb gave rise to the standard sequences of *D. buzzatii* and *D. serido* by inversions 5g and $2j^9$ respectively. After this divergence, *D. buzzatii* has undergone a series of inversions (Wasserman, 1962; Ruiz and Fontdevila, 1981; Fontdevila *et al.*, 1981, 1982). Some of these arrangements (2ST, 2j, $2jz^3$) are ubiquitous, whereas others are restricted to one ($2jq^7$, $2y^3$, $2jc^9$) or few populations (4s) of the ancestral region of the species. This species has colonized in historical times

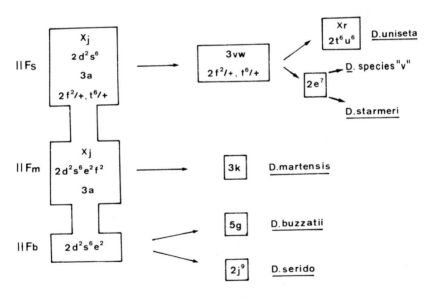

FIGURE 3. *Diagram of the cytological evolution of the South American clusters of the mulleri complex. PRIMITIVE II is subdivided in three: IIFs gave rise to D. uniseta, D. starmeri and D. species "v", IIFm which gave rise to D. martensis and shares 3a with IIFs and $2e^2$ with the buzzatii cluster species, which arose from IIFb. (From Ruiz et al., 1982).*

several areas of the Old World where its chromosomal polymorphism has experienced some interesting qualitative and
quantitative changes. These are relevant for understanding
the colonizing processes (Fontdevila *et al.*, 1982). *D.
serido* from north western Argentina also has incorporated a
series of inversions during its evolutionary history. Ruiz
et al. (1982) have discussed the chromosomal phylads found in
several populations of Argentina ($2j^9 19m^9 n^9$ and $2j^9 k9$) and
the biogeographical differentiation of the polymorphism of
D. serido.

Ancestor IIFm must have evolved into *D. martensis* through
the fixation of 3k. It differs from IIFb, the standard
sequence of the *buzzatii* cluster, by the incorporation of 3a,
$2f^2$ and Xj. These inversions are shared with IIFs, the ancestor of *D. starmeri*, *D. uniseta* and *D.* species "v". Most
of the populations of *D. martensis* studied so far are polymorphic only for inversion $2g^2$, but the population of Cata
(locality 5) shows two additional paracentric inversions
($2o^9$ and $2p^9$) associated with $2g^2$ (Ruiz and Fontdevila, 1982).

Ancestor IIFs must have been polymorphic for $2f^2$ and $2t^6$
and remained so in the next step when through the fixation of
3vw, it gave rise to the ancestral population of *D. starmeri*,
D. species "v" and *D. uniseta*. This last species is virtually monomorphic and only in a few populations do two arrangements of chromosome 2 (ST and v^6) coexist in significant
frequencies. The species came into existence as an offshoot
of the ancestral population by fixation of the Xr and $2t^6 u^6$
arrangements. The origin of *D. starmeri* is more controversial. This species has retained a high chromosomal polymorphism in the great majority of the populations studied
(Table I). As $2e^7$ is ubiquitous and not shared with *D.
uniseta*, this inversion probably appeared after *D. starmeri*
and *D. uniseta* separated, but prior to the formation of subspecies within *D. starmeri*. *D.* species "v" has fixed $2e^7$ in
its standard sequence so it seems that this species originated before geographical differentiation had occurred in
D. starmeri and after *D. uniseta* split off.

II The Cactophilic Niche

A. *The Trophic Niche of D. buzzatii and D. serido*

Early studies of *D. buzzatii* (Patterson and Stone, 1952;
Carson and Wasserman, 1965) reported that this species was
always collected in the vicinity of *Opuntia* spp. In some

cases, *D. buzzatii* was raised from rotting fruits of *Opuntia ficus-indica* (Carson and Wasserman, 1965), but it was not until much later that several authors (Barker and Mulley, 1976; Fontdevila *et al.*, 1981) observed unequivocally that rots of cladodes of several species of *Opuntia (O. ficus-indica, O. maxima, O. quimilo, O. monacantha)* contained larvae and yielded adults of *D. buzzatii*. Evolutionary and historical evidence support the idea that *D. buzzatii* originated somewhere in the Argentinian Monte, perhaps in an area close to the Sierra de San Luis (localities 3, 4 and 10, see Fig. 2). From there it was able to colonize extensive areas by following *Opuntia* species that had been spread by human activities. At present *D. buzzatii* is found in most, but not all, localities where *Opuntia* occurs in the Palearctic, Ethiopian, Australian and Neotropical regions (Carson and Wasserman, 1965; Barker and Mulley, 1976; Fontdevila *et al.*, 1981, 1982), but it is absent from the Nearctic region. The reasons why this species has been so successful in colonization are presently under investigation (Fontdevila *et al.*, 1981, 1982). These studies have shown so far that the chromosomal changes in colonization are twofold. First, chromosomal polymorphism has remained high in the original areas of colonization and has been reduced by founder effects in secondary regions of colonization. Second, arrangement frequencies have changed. Current investigations show that changes in the second chromosome arrangement frequencies of experimental populations are dependent on the feeding substrates: *Opuntia* cladodes and fruits (Ruiz and Fontdevila, in preparation). We are still far from knowing whether these changes are correlated with the chemistry of the rotting cladode and fruit, but there are some indications that the type and concentrations of alcohol may play some role (Fontdevila, 1980).

The role of yeasts in the adaptation of *D. buzzatii* to the cactophilic niche is most controversial. It has been recognized that yeasts are a substantial component of natural dietary intake by *Drosophila* (Heed *et al.*, 1976; Starmer *et al.*, 1976). The differential effect of yeasts on fitness also has been observed several times for various species of *Drosophila* (Wagner, 1949; Barker, 1977). Some of the yeasts present in the rots of two of the most important host plants for *D. buzzatii (O. ficus-indica* and *O. maxima)* have been tested for differential viability by my student F. Peris. Viability differences are highly significant among yeasts and *Pichia cactophila*, an ubiquitous species, shows the strongest negative effect on the development of *D. buzzatii* larvae. This is shown by their low viability and by the lengthening

of their developmental time.   These experiments were performed with yeast-malt extract cultures and, apparently, the
rank performance of *P. cactophila* as a *Drosophila* nutrient
can be improved when cactus extract is added to the culture
(Vacek, 1982).

   The comparative study of the niches of *D. buzzatii* and
D. serido also may provide an interesting example of how
adult host selection is an intriguing component of total
fitness.   Emergences from rots of *O. quimilo* produce only
D. buzzatii, suggesting an initial assumption that some columnar cacti sympatric with *O. quimilo* may be the natural host
of *D. serido*.   However, my student H. Naveira demonstrated
that larvae of *D. serido* can reach the adult stage in *O.
ficus-indica* rots.   The emerged adults were able to reproduce
and females laid eggs that developed normally in *Opuntia*
media.

   These results suggested that adult selection should be
the crucial step of niche differentiation between species.
An experiment on survival time showed that *D. serido* adults
kept on *Opuntia* rots had a short mean life, not significantly
different from that of adults kept in wet cotton vials (control).   On the other hand, *D. buzzatii* adults lived much
longer on *Opuntia* rots than in the controls (Fig. 4).   The
conclusion is that *D. serido* adults are unable to utilize
Opuntia rots as a feeding substrate.   They die before maturity, thus explaining why I was unable to keep a *D. serido*
population on *Opuntia* as the sole feeding source.   We do not
know yet what are the crucial steps in the trophic divergence
of the pair *buzzatii-serido*, but in this case adult selection
may play a critical role and deserves to be investigated.

B. *The Cactophilic Niche of the martensis Cluster Species*

   The niche definition of the *martensis* species is still in
a preliminary phase.   Professor M. Benado and his associates
G. Garcia and H. Cerda at the Universidad Simón Bolivar
(Venezuela), in cooperation with my research group, have
analyzed the emergences from a total of 142 rots taken to the
laboratory from several localities.   The main conclusions
from this study, (Benado *et al.*, in preparation) may be summarized as follows: *D. martensis* is the species with the
widest cactus niche.   It utilizes *Ritterocereus griseus,
Subpilocereus repandus, Acanthocereus* sp. and *O. elatior*.
On the other hand, *D.* species "v" shows the narrowest cactophilic niche.   It utilizes *Opuntia elatior* as its only cactus feeding substrate, confirming an earlier suggestion made

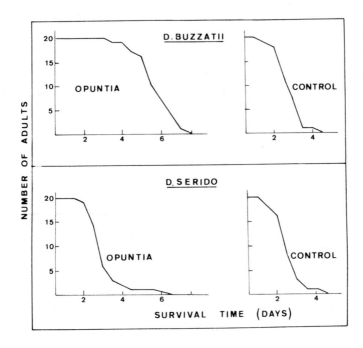

FIGURE 4. D. buzzatii and D. serido survival.

by Ruiz and Fontdevila (1982). *D. uniseta* utilizes *R. griseus* as its most common host (Benado *et al.*, 1979) and *S. repandus* only occasionally.

The niche of *D. starmeri* is more controversial. We have obtained emergences from a few of the *O. wentiana* rots and also from *S. repandus*. These results indicate that *D. starmeri* has a versatile niche, being able to switch from opuntioid to cereoid cacti according to circumstances.

The situation appears to be complicated, but some generalizations can be advanced. In particular *D.* species "v" is chromosomally, morphologically and reproductively very close to *D. starmeri*, but shows a clear niche displacement from that species. The simplest explanation is that a fraction of the *D. starmeri* species became geographically isolated in some arid zone of Oriente during a period of contraction of the deserts. This small fraction converged somewhat towards *D.* species "v", but the final steps of differentiation were given when a new expansion of the deserts produced a secondary contact between the incipient *D.* species "v" and the bulk of the gene pool of *D. starmeri*. This induced a kind of niche displacement of *D.* species "v" in the direction of

O. elatior, which may act as an ecological refuge. This hypothesis rests on several assumptions which remain to be proved, in particular, that the *D. starmeri* strategy is indeed very flexible.

III Inversions and Speciation: A New Approach to the Evolution of Reproductive Isolation

Several authors, through an inclusive analysis of those cases where specific cytological differences are not correlated with the degree of evolutionary divergence (Wasserman, 1982b) or where the reproductive isolation factors are not located primarily in those chromosomes which show the greatest evolutionary dynamics for inversions (Zouros, 1981) conclude that in *Drosophila*, cytological evolution seems to be independent in great part from the fundamental steps of speciation.

The genetic basis of sterility in hybrid males has been studied in a preliminary way by my student H. Naveira, using a planned scheme of successive backcrosses between female interspecific hybrids (*D. buzzatii* males x *D. serido* females) and *D. buzzatii* males. This series of crosses led to the isolation of several *D. buzzatii* strains each of which was introgressed with different chromosomal segments of *D. serido* in the heterozygous condition.

These segmental strains were checked each generation by hybrid female selection using allozyme and chromosomal markers. The sterility test was performed by crossing one male of each hybrid strain with two or three virgin females of *D. buzzatii*. Sterile male recognition was done by using diagnostic allozyme markers (*Odh, Lap*) in heterokaryotypes of the second chromosome. In the case of other chromosomes, indirect tests were used.

The results (Table IV) indicate that all chromosomes contain sterility factors, but that they are not distributed at random. In particular, when the segmental hybrid includes an autosomal inversion of *D. serido*, except in the case of 21^9, male sterility is total. On the other hand, when the hybrid segment does not belong to an autosomal inverted segment, the males are not sterile. This is the case of segments 2(A-B3b) and 5(A-B3d), produced by recombination and detected by observation of segmental asynapsis. Sterility factors in the X chromosome produce lethality in X(A-C2d) hybrids and sterility in X(A-A4e;C1a-H) hybrids.

TABLE IV. *Analysis of Male Sterility in Segmental Hybrids for Several Chromosomes of D. serido Introgressed into D. buzzatii*

D. serido chromosomal segment in heterozygosis	Number of tested males	Number of sterile males	Number of fertile hybrid males	Estimated number of hybrid males
2 $j^9 l^9 m^9 n^9$	92	35	0	35
2 $j^9 k^9$	38	13	0	13
2 $j^9 m^9 n^9$	46	21	0	21
2 l^9	40	2	17	17
2 (A-B3b)	35	0	17	17
3 k^2	176	78	0	78
4 m	124	63	0	63
5 w	139	62	0	62
5 (A-B3d)	48	0	21	21
X (A-C2d)	27	*Male hybrids not viable*		
X (A-4e; Cla-H)	22	12	0	12

A possible role of some inversions in speciation could be as putative protectors during secondary contact of the previous genetic trophic adaptation reached in allopatry. Otherwise, this primary divergence would be swamped out by hybrid recombination. This speculative hypothesis rests on the assumption that some chromosomal segments would include trophic and sterility factors in close linkage. The other serious objection to the model is the relatively low frequency of production of chromosomal rearrangements in natural populations. The final outcome could be the fixation of different arrangements in each of the splitting species, after a phase of transient polymorphism with negative heterosis.

The second chromosome is the most polymorphic in the whole *repleta* group. Since most of the evolution of the group has been produced as a response to the challenge of new available cactus substrates, the evolutionary dynamics of the second chromosome would suggest a leading role in trophic adaptation. At least in *D. buzzatii* we have found some evidence in this respect. If second chromosome inversions are also concerned with hybrid sterility, trophic and sterility factors may be linked in inversions. The frequency of generation of inversions may be enhanced in hybrids, as during these experiments, new inversions and translocations

were observed at a frequency higher than normal. Other unusual phenomena were also detected, such as female sterility.
This set of observations may be related to the syndrome known
as "hybrid dysgenesis" (Sved, 1976; Kidwell *et al.*, 1977).
  These facts suggest that at least some inversions may
have assisted speciation in the secondary contact of subspecies. The identification of these is of crucial importance for understanding evolution.

REFERENCES

Barker, J.S.F., and Mulley, J.C. (1976). *Evolution 30*, 213.
Barker, J.S.F. (1977). *In* "Measuring Selection in Natural
    Populations" (F.B. Christiansen and T. Fenchel, eds.),
    Lecture Notes in Biomathematics *19*, 403. Springer-Verlag,
    Berlin.
Benado, M., Navarro, M.C., and de la Rosa, C. (1979). *Acta
    cient. venez. 30*, 237.
Carson, H.L., and Wasserman, M. (1965). *Am. Nat. 99*, 111.
Crow. J. (1942). *Univ. Texas Publ. 4228*, 53.
Fontdevila, A. (1980). *In* "Ecology and Genetics of Animal
    Speciation" (O.A. Reig, ed.), Equinoccio, Universidad
    Simón Bolivar.
Fontdevila, A., Ruiz, A., Alonso, G., and Ocãna, J. (1981).
    Evolution 35, 148.
Fontdevila, A., Ruiz, A., Ocãna, J., and Alonso, G. (1982).
    Evolution (in press).
Heed, W.B., Starmer, W.T., Miranda, M., Miller, M.W., and
    Phaff, H.J. (1976). *Ecology 57*, 151.
Kidwell, M.G., Kidwell, J.F., and Sved, J.A. (1977). *Genetics
    86*, 813.
Patterson, J.T., and Stone, W.S. (1952). "Evolution in the
    Genus Drosophila." MacMillan Co., New York.
Patterson, J.T., and Wheeler, M.R. (1942). *Univ. Texas Publ.
    4213*, 67.
Ruiz, A., and Fontdevila, A. (1981). *Drosophila Inf. Serv.
    56*, 111.
Ruiz, A., and Fontdevila, A. (1982). *Acta cient. venez.* (in
    press).
Ruiz, A., Fontdevila, A., and Wasserman, M. (1982). *Genetics*
    (in press).
Starmer, W.T., Heed, W.B., Miranda, M., Miller, M.W., and
    Phaff, H.J. (1976). *Microbial Ecology 3*, 11.
Sved, J.A. (1976). *Aust. J. Biol. Sci. 29*, 375.
Vacek, D.C. (1982). Chapter 12, this Volume.

Wagner, R.P. (1949). *Univ. Texas Publ. 4920*, 39.
Wasserman, M. (1962). *Univ. Texas Publ. 6205*, 85.
Wasserman, M. (1982a). Chapter 4, this Volume.
Wasserman, M. (1982b). *In* "The Genetics and Biology of Drosophila." Vol. 3b (M. Ashburner, H.L. Carson, and J.N. Thompson, Jr., eds.), (in press).
Wasserman, M., and Koepfer, H.R. (1979). *Genetics 93*, 935.
Wasserman, M., Koepfer, H.R. and Ward, B.L. (1973). *Ann. ent. Soc. Amer. 66*, 1239.
Zouros, E. (1981). *Genetics 97*, 703.

7

Evolutionary Aspects of Cactus Breeding
Drosophila Species in South America

F. de M. Sene, M. A. Q. R. Pereira,
C. R. Vilela

Instituto de Biociências
Universidade de São Paulo
São Paulo, Brasil

I Introduction

Since 1976, our studies of *repleta* group species in
South America have emphasised genetic differentiation between
populations and modes of speciation. These species inhabit
open formations, and the study area is limited to the west by
the Andes, and to the northwest by the Amazonian Forest. The
main characteristics of this area are the absence of geographical barriers and the tropical climate with relatively
little seasonal variation. The main vegetation domains are
the cerrado, caatingas, Chaco, Atlantic Forest, and
Araucaria Forests, but smaller formations such as dunes,
island forests and highland fields (campo rupestre) also are
included (Ab'Saber, 1977a; Sene *et al.*, 1980).
Palaeoclimatic studies in South America (Ab'Saber, 1977b;
Vanzolini, 1981) show that in the Quaternary period, there
have been recurrent cycles of cold-dry and hot-wet weather.
In the former, vegetation adapted to dry conditions expanded
and the Forests contracted, while in the latter the situation
was reversed. The available data suggest that the period
18,000 to 13,000 years ago was dry, and that the vegetation
distribution was different to that existing at present. In
particular, the cerrados area was reduced and surrounded by
xerophitic formations that linked the areas of the caatingas
and Chaco. The Atlantic and Amazonian forests were reduced
to small islands, and the open formations showed a large

ECOLOGICAL GENETICS AND EVOLUTION
ISBN 0 12 078820 9

expansion along the Atlantic coast, resulting in another link between the caatingas and the Chaco. In the following wet cycle, the cerrados and forests expanded, isolating areas of dry vegetation.

In this large territory from the Atlantic coast of northeastern Brasil through to northern Argentina, the Cactaceae are being used as markers of areas that were dry in the past, as well as at present. Many of the species of *Drosophila* associated with the Cactaceae presumably followed the expansion and contraction of the distribution of the Cactaceae. Thus our aim is to investigate the role of these population expansions and contractions, and the changes in distribution in the evolutionary process.

Some 200 collecting trips have been done, using artificial baits (Sene *et al.*, 1980, 1981) or collecting larvae and adults directly from the breeding sites (Pereira *et al.*, 1982). We summarize here the main results obtained so far.

II Systematic Revision

A general revision of the *repleta* group has been done by Vilela (1981, 1982), using the morphology of male genitalia as the main character for species differentiation. The comparative analysis of male genitalia generally confirmed the phylogenetic relationships established previously by polytene chromosome inversion analysis. Studies of the field collected flies increased the number of species in the *repleta* group recorded from Brasil from the 12 previously noted (Mourão *et al.*, 1965) to 23, with 11 of these being *fasciola* subgroup species. Unlike most *repleta* group species, the *fasciola* subgroup species chiefly inhabit rain forests. One interesting finding concerning this subgroup was the discovery of the breeding site of *D. onca*, on stems of an epiphytic cactus of the genus *Rhipsalis*. Thus this species has retained its fidelity to a cactus host, in spite of being a forest dweller (Sene *et al.*, 1977).

The group revision has added no remarkable new information on the distribution of *mulleri* subgroup species, except for the first record of *D. aldrichi* in South America. It has, however, provided a basis for some speculation on the zoogeographic origins of certain species and suggested more appropriate phylogenetic relationships. These aspects will be discussed separately for each of the species so far studied.

III *Drosophila buzzatii*

The geographical distribution of this species has been
discussed by Vilela *et al*. (1980, 1982), and it has been
reared from *Opuntia vulgaris, O. ficus-indica* and from one
species of *Cereus* cactus, probably *C. peruvianus* (Pereira *et
al*., 1982). It is very abundant in the Chaco, where it is
associated with a number of species of cactus. Outside of
the Chaco, it is rarely found in natural environments, except
for southern Brasil where relatively large populations were
found. In the caatingas areas, it has been collected only in
plantations of *O. ficus-indica*, introduced by man as cattle
food. Even in the continent where it presumably originated,
this species has not expanded its distribution into apparently suitable habitats, except through human intervention in
the movement of cactus.

In the Chaco region, inversion 2J shows a cline from west
to east. In collections from the slopes of the Andes, the
inversion $2JZ^3$ has been found, as well as two new inversions:
$4a^2$, described by Wasserman (1982) as 4s, and $5c^2$. These
chromosomal data will be published, together with allozyme
analyses done by J.S.F. Barker and P.D. East, in the near
future. Outside the Chaco, *D. buzzatii* is practically chromosomally monomorphic.

Analyses of metaphase chromosomes, with special emphasis
on the amount and distribution of constitutive heterochromatin, were done using strains from the different geographical
areas by Baimai *et al*. (1982). There were no differences
among strains, and the metaphase plate was similar to that
previously described by Wharton (1943).

IV *Drosophila meridionalis*

This species has a wide distribution, and has been found
at low frequency in most of the collections (Vilela *et al*.,
1982). The breeding sites identified are *O. ficus-indica,
O. vulgaris, Cereus pernambucensis* and *C. peruvianus*
(Pereira *et al*., 1982). It has been found always in sympatry
with *D. serido*, and on a number of occasions, both species
have been reared from the same stem of rotting cactus collected in nature.

The metaphase chromosome analysis done by Baimai *et al*.
(1982) clearly shows differences among strains from different

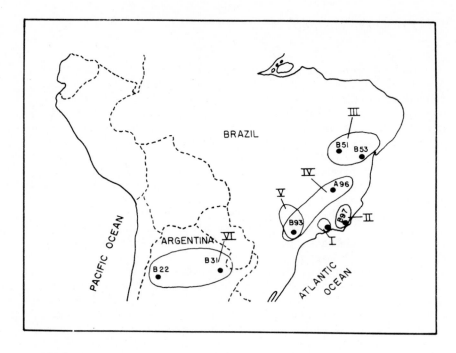

FIGURE 1. *Known geographical distribution of the six different types (I-VI) of metaphase chromosomes of Drosophila meridionalis (After Baimai et al., 1982).*

geographic regions (Fig. 1.).   A standard polytene chromosome strain has been established to allow investigation of the presence or absence of inversions.

V   *Drosophila borborema*

D. borborema is the second most abundant species of the *repleta* group in the caatingas area (*D. serido* being the most abundant), and has been found breeding in *Cephalocereus piauhyensis* (Pereira *et al.*, 1982).   However, it has never been collected outside the caatingas (Vilela *et al.*, 1982).
Because of the difficulty of keeping this species in laboratory culture, we have analysed only male genital morphology.   On this analysis and polytene chromosome analysis (Wasserman, 1982), *D. borborema* is closely related to *D. serido* and to a new and still undescribed species from Puerto Rico.

VI *Drosophila serido*

Most of our work has concentrated on this species, but much remains to be done before the complexity of this apparent species complex can be understood.

Specimens currently identified as *D. serido* are well represented in the collections from most of the areas sampled where cacti are common (Vilela *et al.*, 1980, 1982). However, it is rare in the Chaco, and absent in *O. ficus-indica* plantations. This could explain the fact that *D. serido* has not been transported around the world by man, as has *D. buzzatii*. However, *D. serido* has been reared from *O. ficus-indica*, and from a number of other cactus species, viz., *O. vulgaris*, *Cephalocereus piauhyensis*, *Cereus pernambucensis* and *Cereus* sp. (Pereira *et al.*, 1982).

The geographic populations of this species complex have been partially characterized by studies of male genital morphology, chromosome inversions and metaphase chromosomes.

A. *Analysis of the Morphology of Male Genitalia*

There are at least five types of aedeagus (Silva, 1982), the differences being mainly in the angle of curvature of the dorsal part and in the extent of fused regions. The known geographical distribution of these different types is shown in Fig. 2. Type A, which is similar to that described by Vilela and Sene (1977), is found in the eastern caatingas and in all the areas along the coast. Type B is found in the western caatingas, in the eastern border of the Chaco and in one area of highland (campo rupestre vegetation) in Central Brasil. Type C is found exclusively in areas above 1300 m in the highlands (Espinhaco Range) which divide the caatingas into eastern and western regions. Type D is found in southern Brasil and northeast Argentina, while Type E is found only on the eastern slopes of the Andes in Argentina.

B. *Chromosome Inversions*

Analysis of samples from different populations through comparison of the banding sequences of the polytene chromosomes has been done by Tosi (1982), who gave major attention to chromosome II. The nomenclature used to designate the inversions is not definitive and the system used by Wasserman (1982) will be adopted in future.

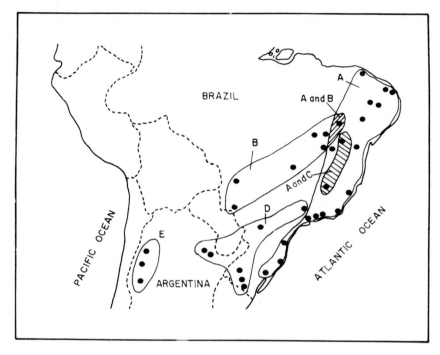

FIGURE 2. *Known geographical distribution of the five different types (A-E) of the aedeagus of Drosophila serido; cross-hatched = overlapping areas (After Silva, 1982).*

The geographical distribution of the inversions is given in Fig. 3. Strain A77 from northeast Brasil has the sequence $2\ a\ b\ m\ n\ z^7$, similar to the ancestor proposed by Wasserman (1982). By comparison with this arrangement, the other strains can be divided into two groups: (1) 2W inversion fixed, and (2) 2Y inversion fixed. Strains of the former group are polymorphic for the 2a inversion. The strains fixed for 2Y show a very wide distribution from northeast Brasil to the slopes of the Andes in Argentina, but can be placed into geographical groups by polymorphism for particular inversions, on other chromosomes as well as chromosome II, as follows:

B58 - 2c/+
A16/B53 - 2b/+, 2c/+
B31, B32 - 2d/+, 2c/+, 5a/+
B20, B25, B26 - 2f/+, 2g/+, 2h/+; 3a/+; 4a/+; 5b/+.

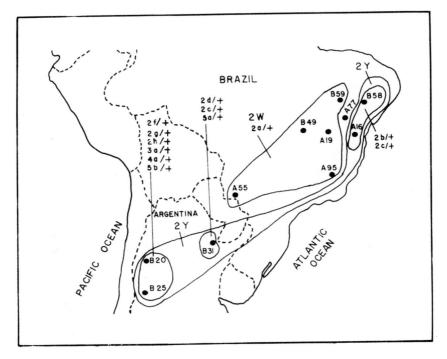

FIGURE 3. Known geographical distribution of several polymorphic and monomorphic polytene chromosome intersions of Drosophila serido (After Tosi, 1982).

C. *Metaphase Chromosomes*

A general survey of metaphase chromosomes (Baimai *et al.*, 1982) showed that strains from different geographic regions could be differentiated on the basis of heterochromatin patterns. The populations of *D. serido* can be divided into six groups on the structure of the microchromosomes ("dots") and the sex chromosomes (Fig. 4).

D. *Synthesis and Further Studies*

Clearly *D. serido* is at least a polytypic species. How- ever, it is still uncertain whether it is just one biological species or a cluster of species. Further polytene chromosome analysis is necessary and in addition, a study of reproduc- tive isolation between strains from different geographic regions has been started. Preliminary results show

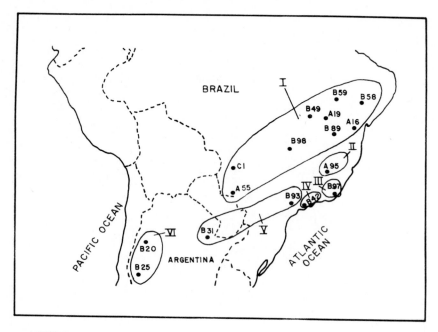

FIGURE 4. *Known geographical distribution of six different types (I-VI) of metaphase chromosomes of Drosophila serido (After Baimai et al., 1982).*

inviability and sterility of F_1 hybrids in some crosses. Also it is known that an insemination reaction is a common phenomenon in the *mulleri* subgroup, and a study of this reaction in homogamic and heterogamic crosses is being done by N.M.V. Bizzo in our laboratory.

Results obtained so far, particularly on metaphase and polytene chromosomes, suggest that the ancestral form of *D. serido* could have originated in northeast Brasil, subsequently expanding to the south and west. The patterns of distribution of the different populations certainly suggest that the evolutionary history of this species complex has been influenced by the palaeoclimatic history of South America.

VII   Conclusions

The results summarized here on the four species of the *repleta* group indicate that their evolutionary histories have been distinct. *D. serido*, *D. meridionalis* and *D. borborema* apparently originated in northern Brasil, and the last species has not spread from this region. On the other hand, *D. buzzatii* most likely originated in the Chaco. However, the data so far available on the geographical distribution and breeding sites, and the analyses of chromosome inversions, metaphase chromosomes and allozyme variation are not sufficient to provide an understanding of its evolutionary history.

For *D. serido*, the differentiation which has occurred in isolated populations is not the same for the different markers studied (Figs. 2, 3 and 4). Thus it would seem that there have been different evolutionary patterns and strategies within this complex, but further study is required before clear conclusions can be drawn.

Nevertheless, it seems that in general the climatic alternations during the Quaternary period have played a very important role in the differentiation of these species and of populations within species.

REFERENCES

Ab'Saber, A.N. (1977a). *Geomorfologia* (Inst. Geogr., Univ. São Paulo) *52*, 21pp.

Ab'Saber, A.N. (1977b). *Paleoclimas* (Inst. Geogr., Univ. São Paulo) *3*, 19pp.

Baimai, V., Sene, F.de M., and Pereira, M.A.Q.R. (1982). (in preparation).

Mourão, C.A., Gallo, A.J., and Bicudo, H.E.M.C. (1965). *Ciênc. Cult., S. Paulo 17*, 577.

Pereira, M.A.Q.R., Vilela, C.R., and Sene, F.de M. (1982). *Ciênc. Cult., S. Paulo* (in press).

Sene, F.de M., Paganelli, C.H.M., Pedroso, L.G., Garcia, E., and Palombo, C.R. (1977). *Ciênc. Cult., S. Paulo* (supl.) *29*, 716.

Sene, F.de M., Val, F.C., Vilela, C.R., and Pereira, M.A.Q.R. (1980). *Papéis Avulsos Zool., S. Paulo 33*, 315.

Sene, F.de M., Pereira, M.A.Q.R., Vilela, C.R., and Bizzo, N.M.V. (1981). *Drosophila Inf. Serv. 56*, 118.

Silva, A.F.G. da (1982). Masters Thesis. Departamento de Biologia, Instituto de Biociências, Univ. São Paulo.

Tosi, D. (1982). Masters Thesis. Departamento de Biologia, Instituto de Biociências, Univ. São Paulo.

Vanzolini, P.E. (1981). *Papéis Avulsos Zool., S. Paulo 34*, 189.

Vilela, C.R. (1981). Ph.D. Thesis. Departamento de Biologia, Instituto de Biociências, Univ. São Paulo.

Vilela, C.R. (1982). *Revta. bras. Entomol.* (in press).
Vilela, C.R., and Sene, F.de M. (1977). *Papéis Avulsos Zool.*, *S. Paulo 30*, 295.
Vilela, C.R., Sene, F.de M., and Pereira, M.A.Q.R. (1980). *Revta. bras. Biol. 40*, 837.
Vilela, C.R., Pereira, M.A.Q.R., and Sene, F.de M. (1982). *Ciênc. Cult., S. Paulo* (in press).
Wasserman, M. (1982). *In* "The Genetics and Biology of Drosophila", Volume 3b (M. Ashburner, H.L. Carson, and J.N. Thompson, Jr., eds.). (in press).
Wharton, L.T. (1943). *Univ. Texas Publ. 4313*, 281.

8
Phyletic Species Packing and the Formation
of Sibling (Cryptic) Species Clusters[1]

R. H. Richardson

Department of Zoology
University of Texas at Austin
Austin, Texas

Crozier (1977) remarked that, "... it would be big news if any widely distributed ant species were *not* to consist of siblings!" "Good" species frequently dissolve into clusters of siblings when studied closely. The pattern of cryptic species clusters is not confined to insects with a social structure. The clues may differ, and the details of speciation may change, but the final pattern seems to be widespread.

In North America the biogeographic distribution reflects several southward progressions of deserts during each ice age, each followed by a northward recession. With each progression species were compressed into the same habitat in a small region in central Mexico, and with each recession some moved northward. The Sierra Madre Occidental in the west and the Sierra Madre Oriental in the east (with the central Chihuahuan Desert) formed a wedge, and most populations moved northward along the coastal plains of the Pacific Ocean and Gulf of Mexico.

The climatological and ecological patterns also propagated a phylogenetic pattern. It is common to find overlapping distributions of very closely related sibling, often cryptic, species of insects sharing habitats. A pattern of speciation may be seen that reflects the compression and

[1]*Work summarized in this report was supported by several grants from the National Science Foundation, National Institutes of Health, and Department of Energy (including AEC and ERDA).*

ECOLOGICAL GENETICS AND EVOLUTION
ISBN 0 12 078820 9

relaxation of species packing in the community, and, with
favorable material can be deciphered into phylogenetic
lineages. However, one must be studying the actual breeding
patterns and the modes of adaptation of a "species" under
natural conditions to detect many of the overlapping ranges
of cryptic species. The *repleta* group of *Drosophila* has been
the most extensively studied insect in this region, and is,
indeed, an effective evolutionary model to guide other
studies, particularily those of insect taxa.

I   History of Systematic Convention in *Drosophila*

The ability to differentiate species is limited by the
taxonomic tools. When the only tools needed to distinguish
species are our eyes, and possibly modest magnification to
see external detail more easily, species are not cryptic.
However, if we need to examine internal anatomy or use
special analytical techniques to differentiate individuals
of different species, the species are called cryptic. Mayr
(1942) used "sibling species" to describe phenotypically
similar, closely related species. Sibling species often are
cryptic, at least until an observant taxonomist notices a
subtle, but consistent difference in morphology.
When Sturtevant (1942) was building the basis for modern
Drosophila systematics, he felt that only species which
could be identified from pinned specimens warranted recognition with an official Latin name. The species thereby
could be recognized by examination of paratypes and holotype
specimens in a museum. However, Wheeler, Patterson, Stone,
Dobzhansky, Hardy, Takada, and other *Drosophila* taxonomists
have placed a greater emphasis on naming species that could
be distinguished by *any* means -- mating tests, internal
anatomy, and genetic analyses were their tools (Patterson
and Stone, 1952). By using this approach they fostered the
studies of speciation, adaptation and biogeography in
Drosophila, as well as other insects, raising the analysis
of evolutionary processes to a new level of resolution. The
identification of incipient species, semispecies, and sibling species became common events, and studying the mechanisms of achieving genetic isolation and measuring genetic
differentiation, with or without comparable morphological
diversity, became the focus of research in evolutionary
genetics. A profound change also occurred when living
"paratype" cultures became important adjuncts to the pre-
served museum specimens. The National Drosophila Species

Resource Center at the University of Texas at Austin is a
direct result of the need for such paratype cultures.

There are numerous examples where sexual isolation is
adequate under natural conditions to allow speciation, but
may be overwhelmed under laboratory tests. Crowding may
prevent mate discrimination, diurnal rhythms may be upset,
or habitat selection may be prevented. Sometimes individuals
of different species extensively hybridize under unnatural
contact. The existence of either reproductive or genetic
isolation in the laboratory generally implies isolation in
nature, but the reverse is definitely not true. Absence of
isolation in the laboratory only indicates that the test
must be conducted in nature, or at least under more natural
conditions.

Initially, genetic isolation was inferred from the
presence of reproductive isolation. Now often it may be
observed directly, even in the absence of marked hybrid
inviability or infertility. We have developed additional
techniques for distinguishing cryptic species in mixtures
collected in the natural habitat (Makela and Richardson,
1977, 1978; Richardson, Ellison and Averhoff, 1982). When
genetically differentiated populations fail to intermate
even though the opportunity exists, there will be fewer
heterozygotes than if they were intermating. Electrophoretically detectable variation usually allows identification of
heterozygotes, as does chromosomal analysis. Therefore, data
from wild individuals derived from electrophoretic and
chromosomal studies can be used to determine the pattern of
mating outside the laboratory.

II Concordance of Phylogenies, Biogeographies and Ecologies

A. *Species Groups*

Throckmorton (1982) outlined the broader scale of evolution in the genus *Drosophila*, and showed that major lineages
each have a characteristic habitat. At times the lineage is
subdivided into Old World and New World portions, but only
one habitat is involved. Major migrations across large land
masses subsequently separated by continential drift did not
materially affect the association between a major lineage and
its habitat. However, major new lineages (species groups)
typically reflect a new habitat invasion. It is especially
significant that the morphological differences that were used
to define taxonomic species groups ultimately have such a
high concordance with habitat similarities.

From Throckmorton's overview of the evolution of the
genus, it is both significant that there were major splits
of lineages associated with habitat shifts, and that these
patterns have remained after millions of years. The separation of major land masses, emergence of mountain ranges,
major climatic changes, formation of island chains and
emergence of other major taxa have not disrupted the initial
direction of evolutionary trends.

Both morphological similarity and habitat similarity are
concordant with phylogenetic relatedness, also determined by
independent criteria, such as chromosome structure. Chromosome diversity between species groups generally is so great
that constructing phylogenies becomes much more difficult.
Nevertheless, there has been progress.

Where polytene chromosome detail is observable, Stalker
(1972) and Yoon, Resch and Wheeler (1972) have connected the
most closely related species groups. Their technique is
based both on maintaining identical banding sequences and on
band morphology across taxonomic gaps. Banding sequences may
be disrupted by chromosomal changes, but common banding
morphology represents similar patterns of gene regulation
since active genes localized in bands often are associated
with "puffs" in the region.

Also DNA sequence comparisons are offering some promise
for phylogenetically relating species groups (Richardson and
Yoon, 1977; Triantaphyllidis and Richardson, 1980;
Richardson, Triantaphyllidis and Turner, unpub.). The
initial studies indicate rates of genetic divergence are
higher in the early stages of the formation of a new species
group, in agreement with the chromosomal studies (Yoon and
Richardson, 1976). A faster rate is probably expected since
a new group is generally exploiting new resources in a new
habitat, and greater genetic change would be anticipated for
adaptation than if a new niche in the same habitat was
involved.

B. Subgroups and Species Complexes

1. The Repleta Group. Heed (1982) has outlined the patterns of biogeographic distributions and habitats of the
repleta group in North America, and finds the patterns still
concordant with phylogenetic relationships and patterns of
chromosome similarity (Wasserman, 1982). There is a characteristic V-shaped distribution, with an occasional type in
central Mexico. The greatest diversity is in southern Mexico,
and the ranges of several species extend north eastward and

north westward. While some types, such as *D. arizonensis*, today are restricted to the northern tips of the V-distribution, they presumably moved northward with the warming following an ice age from a once single population in the south. Those that are more generalist with respect to north-south habitat differences form a complete J- or V-shaped distribution.

Broadly ranging species overlap in the south (the angle of the V) with each other as well as with the more localized species, producing a higher number of species in the *Drosophila* community. This increased number of sympatric species constitutes a greater diversity in this community than those found in central and northern Mexico. Species diversity in this area may have been even greater at the height of the ice ages, when the species at the northern tips of the V-shaped distribution were located more to the south. The present time, between ice ages, may have less species diversity even in southern Mexico than was true for most of the history of this habitat. Most of the time climates were colder than at the present. With the shorter warmer times like the present allowing the spread of species ranges which reduce the overlap, the lowered local species diversity leaves a more loosely packed community, ripe for speciation and phyletic species packing (Richardson and Smouse, 1975).

Thus, the broad pattern outlined for species groups of the genus describes clusters of species within subgroups. Habitats are subdivided within a subgroup, and even a species complex, but still having a high concordance with phylogenetic relatedness. Usually the closely related species can be found exploiting a common taxonomic section of the cacti which share a chemical similarity as well as morphological similarity (Gibson, 1982; Richardson and Smouse, 1976). When several species use a common larval substrate, such as between columnar cacti vs. prickly pear cacti, they also tend to be phylogenetically clustered. The general pattern described by Throckmorton is seen among more closely related species, which have evolved much more recently and in closer proximity than those evolutionary splits producing the lineages of species groups.

2. Hawaiian Drosophila. While it is not directly a focus of this symposium, it may be instructive to briefly compare the patterns of the *repleta* group, evolving primarily on large land masses, with species endemic to the

Hawaiian Archipelago. The large scale biogeographic pattern
is among islands, which have emerged as a result of volcanic
activity. The age of the islands forms a linear sequence
from the older islands in the northwest to the younger in
the southeast portions of the chain (Carson *et al.*, 1970).
Migrations and resulting speciation events have not always
followed the geological ages, and several reverse migrations
have occurred from younger to older islands. Nevertheless,
the endemic *Drosophila* form several species groups, such as
the picture wings and modified mouthparts. They tend to have
different ecological trends concordant with taxonomic
diversity (Yoon, Resch and Wheeler, 1972; Montgomery, 1975;
Heed, 1968). The modes of species recognition create important forces providing the morphological diversity for some
of the lineage names -- visual wing displays in the picture
wings and tactile modes involving mouthpart contact in the
modified mouthpart lineage.

A new lineage arose after the present islands were
formed, probably when Maui Nui was the youngest island (Yoon
and Richardson, 1976). The morphological differentiation was
extensive, initially resulting in its being given generic
status. From chromosomal differentiation we suggested a more
rapid change initially than seen in other endemic lineages,
and it has been moved back into the genus *Drosophila*. In
principle the genetic dynamics leading to new species groups
or genera seem to be no more complex than those leading to
new sibling species clusters. The difference between the
inception of a new species group or genus and a new sibling
species cluster lies in the ecological circumstances.

Thus, the major Hawaiian lineages must have occupied
their characteristic habitat relatively early in the radiation of the genus in the islands, and thereafter each
lineage continued evolving within the framework of a particular ecological theme. Subgroups and complexes of sibling
species have arisen very recently in highly restricted
areas, possibly sympatrically (Richardson, 1974). Thus,
evolutionary patterns in Hawaiian *Drosophila* have developed
under very different circumstances of geographical separation, habitat changes, time scales, and adaptive strategies,
but form a lineage pattern very similar to the *repleta*
group.

C. *Sibling Species Clusters*

1. North American Pattern.   There are about five clusters
of sibling species in the *mulleri* complex which serve as a

model of the patterns we are considering. They are (1) *arizonensis, mojavensis, navajoa;* (2) *aldrichi, mulleri, wheeleri;* (3) *ritae* and an undescribed new species; (4) *longicornis, pachuca, propachuca, desertorum;* (5) *martensis, starmeri, uniseta.* The latter cluster is South American, and growing (Wasserman, 1982; Sene *et al.,* 1982). The other species are not yet defined to a sibling cluster, although they are closely related to the *longicornis* cluster. Heed (1982) has given the geographic distributions. We know of undescribed species that increase this list, since they are cryptic within one of the sibling clusters. These include one in the *longicornis* cluster (Heed, 1982) and a presumptive type related to *D. propachuca* which we have observed (Richardson, unpub.). *D. tira* has been combined with *D. ritae* (Valela, unpub.), but an unnamed sibling species is known.

When we examine the clusters of sibling species within a species complex, we see a continuation of the pattern. For example, the cluster of prickly pear breeding species, *D. longicornis, D. pachuca* and *D. propachuca* are indistinguishable even by internal anatomy, and *D. desertorum* differs significantly only in the male genitalia, yet they all may be cultured from one piece of rotting prickly pear cactus collected in nature (Richardson, Smouse and Richardson, 1977; Richardson, unpub.). They are genetically differentiated in some enzyme systems, and there is some reproductive isolation expressed by sterility of males in certain crosses among the three species, excluding *D. desertorum.* They share many chromosome similarities (Wasserman, 1982). In fact, initially they were mistakenly split, with *D. longicornis* removed from close phylogenetic association (Wasserman, 1962), until their molecular similarities were found by M. E. Richardson (personal communication to M. Wasserman). *D. desertorum* is differentiated by several inversions from the other three species of the cluster.

The most striking difference among these four cryptic species lies in their developmental pattern. The ability to reliably culture *D. propachuca* came only after we discovered that the pupation time was almost twice as long as the other two species, and that it was necessary to wait for the adults to eclose long after all the larvae in a culture vial had pupated. This strongly suggests that these cryptic species, often sharing a common mass of rotting cacti, have diverged in their developmental pattern, thereby reducing niche overlap to some degree. *D. desertorum* is easily cultured for a few generations as an isofemale line, but eventually lines must be mixed to maintain a culture. This species presumably is sensitive to inbreeding depression.

These species culture best on a cactus-supplemented medium. There is not a nutritional requirement; the effects of cactus in the medium is more subtle. While the factors are unknown, our conjecture from the work of Fogleman (1982), Kircher (1982) and Starmer (1982) is that, under natural conditions, these species may feed on different yeasts and exploit slightly different stages of the rotting tissue. Diversity which might separate cryptic species might be expected to have been important adaptations in the evolution of their niche at the time they were speciating.

From southern Mexico most species of the *mulleri* complex have a characteristic V-shaped distribution. The greatest diversity is in southern Mexico, and the ranges of several species extend north eastward or north westward. While they once formed a single population in the south, once separated to the north, genetic divergence commonly results. For example, *D. aldrichi* often shows male sterility of hybrids between strains from Sonora and from Texas, suggesting an incipient species-level diversity. There also is a difference in mating speed between these two subspecies, but no consistent discrimination was shown (Richardson, unpub.). *D. wheeleri*, the closest relative to *D. aldrichi*, seems to have arisen in the northwestern extreme of the range of *D. aldrichi*, and shows a similar pattern of reproductive isolation to both eastern and western subspecies of *D. aldrichi*. There may be a region of sympatry of the eastern and western types of *D. aldrichi* in the state of Hidalgo. If this were studied, it would allow a test for genetic isolation under natural conditions. Should they be isolated in sympatry, we can formally separate *D. aldrichi* into two species, much like *D. arizonensis* and *D. mojavensis* in Sonora.

Even *D. arizonensis* from eastern Mexico inhabits a different rotting cactus than when found in western Mexico, and they are genetically differentiated as well (Richardson, Smouse and Richardson, 1977). There is no detectable reproductive isolation in the laboratory tests. Therefore, we cannot determine whether they are actually different species until they become sympatric and can express any intrinsic genetic isolation.

Adults are physiologically responsive to their microhabitats and have highly differentiated niches comparable to larvae. Eckstrand (1979) examined the ability of several cactus breeding species to physiologically regulate the loss

of water from their body. Most were similar at high humidities, but species from dry habitats were able to regulate
the loss of water at lower humidities. In one case, we found
greater diversity in water balance between two sibling
Hawaiian *Drosophila* species in a relatively moist habitat
(Eckstrand and Richardson, 1981) than between one of them
and *Drosophila* in the desert habitat (Eckstrand, 1979). The
microhabitat difference is the central factor determining
the physiological diversification of sibling species. Of
course, each species is sufficiently mobile to traverse
several microhabitats each day. Habitat selection concen-
trates their exposure in a particular microhabitat. In the
case of the Hawaiian sibling species, one member of the
sibling pair occupies a microhabitat of deep shade, often on
the underside of leaves, while the other's microhabitat is
open shade, and often the individuals are on the upper
(exposed) side of the leaves. The distance between these
microhabitats, however, is only a meter or two (Richardson
and Johnston, 1975). We are still unravelling the microhabitats of desert *Drosophila*, but Johnston and Templeton (1982)
report some new findings relative to dispersal patterns.

   2. *South American Pattern.*   In 1972 when plans were
being made that M. E. Richardson and I visit the University
of Sao Paulo, Brazil, there was some concern expressed when
I indicated a desire to use the *repleta* group as a model.
Based on earlier surveys by Theodosius Dobzhansky and his
colleagues, only those species of the *repleta* group that
were more associated with humans had been found. There was
no evidence that a complex evolutionary process had occurred
such as we had been studying in North America. Nevertheless,
Drs. Edmundo Magalhaes and Crodowaldo Pavan agreed that I
could reexamine some of these species. Dr. Magalhaes returned
and encouraged Dr. Sene and others to work in the *repleta*
group until I arrived in order to establish some stocks and
gain some preliminary experience.
   Soon after we arrived in Sao Paulo early in 1974 Dr.
Sene and I visited the cactus nursery at the University, and
found some rotting Opuntia with empty pupae cases and
larvae. We took this rot into the lab, and reared a new
species from it. Over the next several weeks we collected
flies and rotting cacti in several areas in Brazil. We found
still other new species. We characterized them by differences in male genitalia as well as their allozyme profiles.
They are now known as *D. serido* and *D. borborema*. At this
time it was apparent that there was a lot of exciting work

to be done in the evolutionary study of the *repleta* group in
South America.
   We visited several people who could tell us more about
the paleoclimate and cactus distributions, and we began to
formulate models of how the radiation in South America might
have occurred. Of course, the ideas from studies in North
America greatly influenced our thinking. It is, indeed,
gratifying to have these hopes beginning to become realized
as reported by Sene *et al*. (1982) and Fontdevila (1982).
   The complexity of *D. serido* reported by Sene and
Fontdevila suggests that a cryptic species cluster exists.
From the differences in metaphase chromosomes shown by Sene,
there might be as many as six different species that have
been collected, and most of the continent has not been
collected! Much work remains to organize these species into
more complete phylogenies, but Wasserman (1982) has made an
initial advance, relating these species to others he has
included in the *mulleri* complex. While it now appears that
the radiations in North America and South America were
mostly independent, they are connected phylogenetically.
When the cacti are better known, it may be possible to
determine the relationships of these two radiations in the
mulleri complex, and also relate the origin of this complex
which lives in the cactus habitat to others that are found
in more moist areas, such as the *fasciola* subgroup.

   *3. North American Clusters of Cryptic Species in
Screwworm Flies.* It is noteworthy that *Drosophila* species of
the *repleta* group in general, and the *mulleri* complex in
particular are characterized as "cactus breeders," in parallel to the new species of *Cochliomyia* (Richardson, Ellison
and Averhoff, 1982) which are "warm blooded animal
breeders." In this genus there is comparable morphological
diversity among about nine groups, and possibly almost this
many species, as there is among the four species of the
longicornis cluster. Male genitalia can be used to separate
adults in some cases, but generally the identification
depends on the detailed analysis of the metaphase chromosomes. They share a larval habitat much like the prickly
pear cactus breeding species in that more than one may
inhabit a single wound, and superficially all are in a wound
of a warm blooded mammal. Gassner and Brommel (personal
communication) found important differences in the microbial
community in the wounds correlated with the presence of
certain types of bacteria. More recently, Foss and McDonald
(personal communication) have found electrophoretically

identifiable groups of *Proteus rettgeri* from screwworm
larvae that differ in wounds of sheep from those found
in wounds in either horses or cattle. Microbial diversity
among host species parallels that of the yeasts found in
different cacti (Starmer, 1982). In addition, these types
of screwworms form a biogeographic pattern similar to the
species of the *mulleri* complex -- a V-shaped overall
distribution, with different species tending to be isolated
east to west in the north, but with extensive southern
overlap. Although fertile hybrids may be obtained in the
laboratory, there is no effective gene flow among them in
nature.

III Significance of Cryptic Species

A. *To Speciation and Biogeographic Problems*

Divergence in similar microhabitats appears to be
typical for the cryptic species clusters. The products of
the divergence, the cryptic species, often overlap extensively in their ranges. Superficially these patterns seem to
suggest sympatric speciation is a common mode.
While sympatric speciation may be the mode in many of
these cases, two features must be considered before the case
is firmly made. The species may be ancient, and their place
of origin may be some distance from the present range. A
detailed reconstruction of the habitats in geological history generally becomes necessary to infer the likely ranges
at times when speciation occurred.
The other feature of their ranges requires examination
on a smaller scale -- the "grain" of their habitat. If the
grain is large, populations in one area (grain) may not have
extensive gene flow with populations occupying other grains
for long times. The community composition in different
grains, even with similarity of microhabitats, may differ
and allow divergence, and possible fortuitous genetic isolation may arise. As Templeton (1980) has shown, speciation
may be rapid when there is a major reorganization of the
genome, and speciation may occur in a diverse, large grain
habitat.
There is no easy way to determine if sibling species
arose sympatrically or allopatrically. The weight of the
evidence must be the guiding factor. However, the process of
speciation now appears to be much less time consuming than
it once did, and we know that the effects of genetic inter-

actions, particularily those regulating genes, may have
almost immediate effects that could result in considerable
genetic isolation (e.g., Templeton and Rankin, 1978;
Templeton, 1980). Reproductive isolation (e.g., hybrid inviability) is not as important as once thought for effective
genetic isolation. Therefore, the theoretical feasibility of
sympatric speciation is increasing. In some instances semispecies or sibling species may be of sufficiently recent
origin to allow better evaluations of the roles and circumstances for sympatric and allopatric speciation dynamics.
The ability to detect populations that are sympatric and
exhibiting significant genetic isolation is crucial to the
study of speciation.

B. *To Paleontological Problems*

Speciation requires (1) a potential niche, and (2) a
genetically adapted type isolated from other populations in
the community. To pack another species into a community, a
colonizer may be from outside the habitat (traditional view)
or from speciation inside the habitat ("phyletic species
packing" (Richardson and Smouse, 1975)). In the latter case
the speciation might be sympatric, or allopatric if the
habitat were course grained. The divergence might be behavioral or physiological, for example, and not greatly modify
external morphology, so that eventually the two diverged
populations may come to occupy a single grain and have adequate genetic isolation to be sympatric sibling (and
cryptic) species. Coming from inside the habitat, both
species would be preadapted in many ways to all the grains
(i.e., to the habitat). When a population colonizes a new
grain, it would have a very much greater chance of becoming
established than would an unrelated type, adapted to a different habitat.

Habitat diversity and species packing is the mode of
natural selection acting in species selection (Wright, 1967;
Stanley, 1975). It follows then that punctuated evolution
(Gould and Eldredge, 1977) is the track of potential niches
that were realized, and this track is the latent image of
the habitat diversity written into the fossil record through
the *morphology* of the species. When a habitat is stable,
phyletic species packing can replace lineages that become
extinct, and maintain an appearance of morphological stability (recorded as a single species, but most probably a
cryptic sibling species complex). When the habitat shifts or
suitable new ones form, the phyletic niche track shifts and

punctuates the morphological record. Thereby, one might view punctuated evolution as punctuated habitat diversity.

The examination of species groups in *Drosophila* shows greater morphological diversity associated with major shifts in habitat, and the morphological diversity observed at the level of the speciation process often is very limited. The pattern of morphological diversity more closely reflects the differences between the niches of the species than the actual incidence of speciation. As other taxa have been examined, a similar pattern is revealed, although there may be differences between taxa in the morphological expression of niche diversity.

The ice ages almost certainly contributed to the dynamics of speciation in the *repleta* group. If the rate of speciation was high when the deserts were expanding as the climate warmed, and the community was becoming more subject to invasion ("loosely packed"), then these were punctuations in the phyletic history. However, the punctuations related to such activity did not necessarily (nor even usually) appear with sufficient morphological differentiation to be detectable if one were examining a fossil record. Furthermore, the origin of the endemic groups of Hawaiian *Drosophila* would be observable if fossilized, but the rates were probably determined by the geological and botanical dynamics of the islands -- punctuations of the *Drosophila* habitat. The morphological diversity is not proportional to the genetic diversity, and the evolutionary dynamics are not proportionally reflected in the morphological pattern.

Speciation in times of greater environmental stability may be at a comparable rate, but with less chance of being reflected in the fossil record. It is impossible among related taxa to separate roles of genetic variation and mechanisms of adaptation from the historical sequence of habitat changes reflected by the niches that were developed during the successful exploitation of the resources by the lineages. This evolutionary conundrum is shared by somatic development, and also has created extensive confusion. "Nature" cannot be separated from "nurture" in the differentiation of cell lineages (nor IQ, nor other phenotype) in the development of an individual any better than can the nature of the genetic system be separated from the nurture of ecological diversity in morphological differentiation of phylogenetic lines.

C. To Ecological and Economic Problems

I once had a vertebrate ecologist tell me that he did
not think that identifying cryptic species was an important
problem in biology, particularily since he had not had to
contend with any in his work with lizards. Another time I
spent several hours with another vertebrate biologist with a
different point of view. I was intrigued by his description
of watching a slow progression of the parthenogenetic form
of a species of lizard move across a region of Brazil. It
was interesting how he related collecting with a slingshot
while waiting for an airplane at a remote airport, and from
one year to another the sexual form had been replaced by the
asexual form. The sex ratio, of 1:1 for the sexual forms and
only females in the asexual form, was the conspicuous feature of change that alerted him to the diversity. The
asexual and sexual forms are cryptic species because individual females could not be categorized without examination
of their mode of reproduction, such as by a progeny test. I
think the difference between these two biologists' perspectives on the genetic diversity is significant. One person
considered genetic diversity more of a nuisance and the
breeding details as inconsequential. The competitive interactions among clearly different (morphologically distinct)
species was the focus of his attention. The other person
was, among other things, focusing on speciation and biogeography.

We have been working on insects recently with much the
same difference in perspective between ourselves and our
associates in the U.S. Department of Agriculture.
Cochliomyia hominivorax has been considered to be one
species throughout the Western Hemisphere. Memory is short.
A few decades ago the saprophytic blowfly, *C. macellaria*,
was considered the same species as the parasitic form. It is
now important to recognize the presence of isolation among
wild types, since the parasitic form is being eradicated
from most of Mexico by the release of sterile flies. While
some behavioral isolating barriers may be overwhelmed by
releasing more sterile flies, it is inefficient (and thereby
very expensive), and some matching between natural populations and released sterile flies may be necessary to do a
complete job.

The taxonomic recognition of cryptic species thereby
becomes critical for identifying differences in habitat
selection, measuring species diversity, and for outlining
geographic ranges. One can argue that logically the evaluation of competitive relations also depends on a full ac-

counting of all species, since we know there are important differences among the *Drosophila* species we have studied. While most ecological studies do not address the problem of counting cryptic species in a community, it seems important in most cases to be sure the community composition is truly as described on the species checklist.

In the case of screwworms, preliminary results clearly show how elimination of one member of a species complex "releases" others (Richardson, Ellison and Averhoff, 1982). Although the competitive interactions are unknown, two species increased dramatically in relative frequency when the most common species was eliminated. The net result was little or no change in parasite load, but the community complexity was reduced by one species.

The taxonomy, based on a biological or evolutionary species concept, must be accurate for many biological pest control programs. The work on screwworms is but one example. Malaria control in Europe was dependent on the recognition of cryptic species of mosquitos, only some of which were vectors. Pheromone attractants are effective only on some types (actually different species) of moths. The list is extensive and growing. The patterns we expect from our studies of *Drosophila* predict that simple monospecific communities based on the external morphology are typical only in recently colonized areas, and the general pattern is for there to be clusters of closely related, often cryptic, species in any "mature" population. While such cryptic diversity makes the systematics more challenging and the study of evolutionary processes more exciting, the problems of pest control are greatly complicated and such diversity is potentially the bane of pest management.

ACKNOWLEDGMENTS

W. B. Heed first introduced me to the ecology of *Drosophila*, and Lynn Throckmorton first introduced me to the *repleta* group. The Hawaiian *Drosophila* became a subject of my attention with the assistance of W. S. Stone, M. R. Wheeler, H. T. Spieth, Elmo Hardy, H. L. Carson and others of the Hawaiian Drosophila Project. These and many others have contributed significantly to the ideas presented here. Most notable are my close associates, J. R. Ellison, W. W. Averhoff and D. Vasco.

REFERENCES

Carson, H. L., Hardy, D. E., Spieth, H. T., and Stone, W. S. (1970). *In* "Essays in Evolution and Genetics in Honor of Theodosius Dobzhansky" (M. K. Hecht and W. C. Steere, eds.), p. 437. Appleton-Century-Crofts, New York.

Crozier, R. H. (1977). *Annu. Rev. Entomol. 22*, 263.

Eckstrand, I. A. (1979). Ph.D. Dissertation, Univ. of Texas at Austin.

Eckstrand, I. A., and Richardson, R. H. (1981). *Oecologia 50*, 337.

Fogleman, J. C. (1982). Chapter 13, this Volume,

Fontdevila, A. (1982). Chapter 6, this Volume,

Gibson, A. C. (1982). Chapter 1, this Volume,

Gould, S. J., and Eldredge, N. (1977). *Paleobiology 3*, 115.

Heed, W. B. (1968). *Univ. Texas Publ. 6818*, 387.

Heed, W. B. (1982). Chapter 5, this Volume.

Johnston, J. S., and Templeton, A. R. (1982). Chapter 16, this Volume.

Kircher, H. W. (1982). Chapter 10, this Volume.

Makela, M. E., and Richardson, R. H. (1977). *Genetics 86*, 665.

Makela, M. E., and Richardson, R. H. (1978). *In* "The Screwworm Problem, Evolution of Resistance to Biological Control" (R. H. Richardson, ed.), p. 49. Univ. Texas Press, Austin.

Mayr, E. (1942). "Systematics and the Origin of Species." Columbia Univ. Press, New York.

Montgomery, S. L. (1975). *Proc. Hawaiian Entomol. Soc. 22*, 65.

Patterson, J. T., and Stone, W. S. (1952). "Evolution in the Genus Drosophila." The Macmillan Co., New York.

Richardson, R. H. (1974). *In* "Genetic Mechanisms of Speciation in Insects" (M. J. D. White, ed.), p. 140. Australia and New Zealand Book Co., Sydney.

Richardson, R. H., and Johnston, J. S. (1975). *Oecologia 21*, 193.

Richardson, R. H., and Smouse, P. E. (1975). *Oecologia 22*, 1.

Richardson, R. H., and Smouse, P. E. (1976). *Biochem. Genet. 14*, 447.

Richardson, R. H., and Yoon, J. S. (1977). *Genetics 86*, s51.

Richardson, R. H., Smouse, P. E., and Richardson, M. E. (1977). *Genetics 85*, 141.

Richardson, R. H., Ellison, J. R., and Averhoff, W. W. (1982). *Science 215*, 361.

Sene, F. de M., Pereira, M. A. Q. R., and Vilela, C. R. (1982). Chapter 7, this Volume.

Stalker, H. D. (1972). *Genetics 70*, 457.

Stanley, S. M. (1975). *Proc. natn. Acad. Sci. U.S.A. 72*, 646.

Starmer, W. T. (1982). Chapter 11, this Volume.

Sturtevant, A. H. (1942). *Univ. Texas Publ. 4213*, 5.

Templeton, A. R. (1980). *Genetics 94*, 1011.

Templeton, A. R., and Rankin, M. A. (1978). *In* "The Screwworm Problem, Evolution of Resistance to Biological Control" (R. H. Richardson, ed.), p. 83. Univ. Texas Press, Austin.

Throckmorton, L. H. (1982). Chapter 3, this Volume.

Triantaphyllidis, C. D., and Richardson, R. H. (1980). *Sci. Annuals, Fac. Phys. & Mathem., Univ. Thessaloniki 20a*, 93.

Wasserman, M. (1962). *Univ. Texas Publ. 6205*, 85.

Wasserman, M. (1982). Chapter 4, this Volume.

Wright, S. (1967). *In* "Mathematical Challenges to the Neo-Darwinian Interpretation of Evolution" (P. S. Moorhead and M. M. Kaplan, eds.), p. 117. Wistar Inst. Press, Philadelphia.

Yoon, J. S., Resch, K., and Wheeler, M. R. (1972). *Genetics 71*, 477.

Yoon, J. S., and Richardson, R. H. (1976). *Genetics 83*, 827.

PART III
YEASTS–CHEMISTRY–*DROSOPHILA* INTERACTIONS

9
Taxonomy and Evolution of Some Ascomycetous Cactophilic Yeasts

D. L. Holzschu[1]
H. J. Phaff

Department of Food Science and Technology
University of California
Davis, California

I Introduction

Studies of a large number of natural isolates of
cactophilic yeasts (i.e., yeasts found only in association
with the necrotic tissues of various cacti) belonging to the
genus *Pichia* have provided information pertaining to their
physiological phenotypes, pattern of host plant distribution,
interfertility, and DNA base composition values (Table I).
Each of the yeast species included in this study is
phenotypically congruous with *P. membranaefaciens* as defined
by Kreger-van Rij (1970), but differs significantly in DNA
base composition from *P. membranaefaciens*; they are not
interfertile with this species or among themselves. Each
species is capable of utilizing glucose, glycerol, ethanol,
succinate, and lactate. Strains capable of utilizing D-
xylose have been identified as *P. heedii*; those requiring a
sulfur-containing amino acid for growth and utilizing
glucono-δ-lactone or K-gluconate as *P. amethionina*; those
growing on salicin and/or cellobiose as *P. opuntiae*; and
those utilizing glucosamine strongly as *P. cactophila*
(summarized in Starmer *et al.*, 1980). Based on this

[1]*Present address: Department of Bacteriology, University
of California, Davis, California*

ECOLOGICAL GENETICS AND EVOLUTION
ISBN 0 12 078820 9

Table I.   Comparison of the intraspecific characteristics of three cactus-specific yeasts with the relative degree of host plant divergence

Organism	Intraspecific Physiological Differences	Intraspecific Conjugation (Spore Viability)	mol% G+C	Taxonomic Position of Host Plants
Pichia heedii	citrate$^+$	good (97.2%)	32.4	Species: Senita, Subtribe Pachycereinae
Pichia heedii	citrate$^-$			Saguaro
Pichia amethionina var. amethionina	mannitol$^-$ ttg^{+a}	fair (0)	33.1	Subtribes: Stenocereinae
Pichia amethionina var. pachycereana	mannitol$^+$ ttg$^-$			Pachycereinae
Pichia opuntiae var. opuntiae	citrate$^+$ ttg$^-$ cellobiose L^b/– mannitol L/– 37o$^-$	poor (asci are rarely formed)	34.0	Tribes: Opuntieae
Pichia opuntiae var. thermotolerans	citrate$^-$ ttg$^+$ cellobiose$^+$ mannitol L 37o$^+$			Pachycereeae (Subtr. Pachycereinae)
Pichia cactophila	none	not studied due to homothallism	36.3	all species of cacti

^{a}ttg = triterpene glycoside resistance
^{b}L = Latent growth

information, the heterothallic strains of *P. heedii*, of *P. amethionina*, and of *P. opuntiae* appear to be at different stages of speciation (Starmer *et al.*, 1980). Strains of *P. cactophila* are homothallic preventing analysis of mating compatibilities among strains and are metabolically quite homogeneous, although occasionally some differences in the ability to assimilate citrate are noted. Most strains form two spores per ascus. The few four-spored strains isolated were shown to be heterothallic (Starmer *et al.*, 1978a).

The biological system from which these yeasts have been recovered is apparently tripartite. Cacti serve as the host plant for cactophilic *Drosophila* and yeasts, with the *Drosophila* presumably the vector for yeast dispersal. In turn, the yeasts provide nutrients for *Drosophila* development. The *Drosophila* species and yeasts do not inhabit healthy cactus tissues but rather succulent tissues that have been exposed due to injury. Soft rot, pectinolytic bacteria initially colonize the tissue followed by the yeasts. The yeasts present are determined by host plant chemistry, attractiveness to vector, competition, and fitness. Yeast and *Drosophila* distribution among host plants are summarized in Figure 1.

These data, which indicate that some of these yeasts may be interrelated, have been used in this study as a natural framework for the formulation of a series of experiments examining macromolecular divergence (as measured by DNA–DNA homology values) among strains of *P. heedii, P. amethionina, P. opuntiae*, and *P. cactophila*.

II Materials and Methods

The yeast strains used in this study were obtained from the yeast culture collection of the Department of Food Science and Technology, University of California, Davis. All strains had been isolated from naturally occurring cactus rots during ecological surveys (Phaff *et al.*, 1978; Starmer *et al.*, 1978a; Starmer *et al.*, 1978b; Starmer *et al.*, 1979; Starmer *et al.*, 1982). Standard tests for identification of yeasts were done as described by van der Walt (1970). The procedures for growing test yeasts in adequate amounts for DNA isolation and purification and for determining DNA base composition values (mol% G+C) by buoyant density equilibrium centrifugation in CsCl have been described by Price *et al.* (1978). *In vitro* labeling of probe DNA with ^{125}I was done as

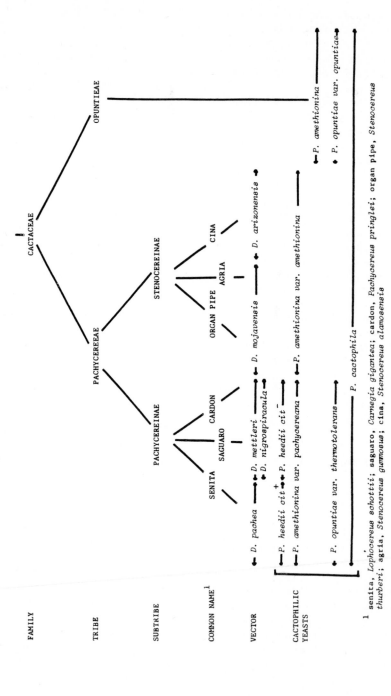

Figure 1. *Cactus phylogeny (sensu Gibson and Horak, 1978); Drosophila and yeast distribution among cacti.*

[1] senita, *Lophocereus schottii*; saguaro, *Carnegia gigantea*; cardon, *Pachycereus pringlei*; organ pipe, *Stenocereus thurberi*; agria, *Stenocereus gummosus*; cina, *Stenocereus alamosensis*

described by Holzschu *et al.* (1979). Procedures for DNA fragmentation, removal of rapidly renaturing sequences from radioactive reference DNA, and DNA reassociation experiments in solution, including renaturation kinetics of homologous DNA fragments, have been described by Price *et al.* (1978).

III Results and Discussion

A. *Pichia heedii*

Pichia heedii, a heterothallic diploid yeast, is isolated almost exclusively from *Lophocereus schottii* (senita) and *Carnegia gigantea* (saguaro), members of the subtribe Pachycereinae (Table I). The strains are metabolically homogeneous with the exception of the utilization of citrate which is variable. Strains isolated from senita utilize citrate while those isolated from saguaro do not. This difference is due to a change at a single genetic locus and could be in response to the chemistry of the host plants (Starmer *et al.*, 1980). Mating experiments among the two metabolic types have shown the strains to be completely compatible with about 95% of the spores from hybrids being viable. Presumably, the two *Drosophila* species that utilize senita (*D. pachea*) and saguaro (*D. nigrospiracula*) maintain the partition of these metabolic subpopulations. *P. heedii* may represent the initial stage of host race formation (Starmer *et al.*, 1980).

Table II presents data in which DNA isolated from the type culture of *P. heedii* (76-356) was used as the reference. Strains used in this experiment were chosen to represent as wide a geographic range as available, i.e. Baja California Sur, Mexico to Sonora, Mexico to Tucson, Arizona, as well as metabolic and host plant diversity (see Phaff *et al.*, 1978, for exact locations of isolation). The DNA isolated from each of the strains of *P. heedii* showed greater than 95% sequence complementarity with the reference DNA, indicating homogeneity within this species. Evidently, the metabolic types (cit$^+$, cit$^-$) of *P. heedii* are not discernible by DNA-DNA complementarity analysis. These findings agree with earlier data: relative similarity of phenotype, relationship of the cactus host plants, and successful matings in the laboratory that suggested a very close relationship among the metabolic types of *P. heedii* (Starmer *et al.*, 1980). Whether or not these populations are genetically isolated but still very similar cannot be deduced from these data. The type

culture of *P. opuntiae* var. *thermotolerans* showed only a low
degree of homology, indicating a relationship to *P. heedii*
that is beyond the resolution of this experiment.   The higher
values of homology with *P. amethionina* var. *amethionina* and
P. cactophila are also considered insignificant in view of
reports by Anderson and Ordal (1972) and Wilson *et al.* (1977)
whose data suggest that DNA complementarity values near 20%
do not correlate with other criteria commonly used for
measuring relatedness among organisms.   This level of
homology between *P. heedii* and *P. amethionina* var.
amethionina and *P. cactophila* was not observed consistently.

Table II.   *Reassocation of ^{125}I-labeled DNA isolated from Pichia
heedii 76-356 with DNA of cactophilic yeast isolates*

Organism or tissue	Isolate number	% Actual binding ± SD[a]	% Relative binding[b]
P. heedii[d]	76-356[c]	80.5 ± 0.7	100
P. heedii[d]	76-280C	79.9 ± 0.7	99.3
P. heedii[d]	72-103	77.6 ± 0.1	96.2
P. heedii[e]	76-295	79.4 ± 0.9	98.6
P. heedii[f]	77-480	79.4 ± 1.2	98.5
P. heedii[f]	73-54	78.2 ± 1.9	96.9
P. heedii[f]	72-114	77.4 ± 0.6	95.6
P. opuntiae var. thermotolerans	76-211[c]	9.8 ± 0.5	5.2
P. amethionina var. pachycereana	76-384A[c]	15.3 ± 1.7	12.6
P. cactophila	76-243A[c]	18.2 ± 1.7	16.5
Calf thymus DNA		7.8 ± 1.6	2.6

[a]*SD = standard deviation; average of triplicate tests, corrected
for zero-time binding.*

[b]*Corrected for self-reassociation of labeled DNA.*

[c]*Type strains*

[d]*Utilizes citrate; isolated from senita.*

[e]*Utilizes citrate; isolated from agria.*

[f]*Does not utilize citrate; isolated from saguaro.*

B. *Pichia amethionina*

Pichia amethionina has two described varieties, *P. amethionina* var. *amethionina* and *P. amethionina* var. *pachycereana* (Starmer *et al.*, 1978b). Notably, this species is a naturally occurring auxotrophic taxon requiring a sulfur-containing amino acid for growth. Strains of *P. amethionina* var. *amethionina* have been isolated from nature as heterothallic diploids, homothallic diploids (Australian strains), and heterothallic haploids (Starmer *et al.* 1978b; W.T. Starmer, personal communication). Most strains of *P. amethionina* var. *pachycereana* were isolated as heterothallic diploid or haploid yeasts, but recently homothallic diploid strains have been isolated from *Opuntia* cacti (W.T. Starmer, personal communication). The first mentioned variety has been isolated predominantly from cacti of the subtribe Stenocereinae and the latter from the subtribe Pachycereinae. *P. amethionina* var. *amethionina* is unable to utilize mannitol (man$^-$) as a sole source of carbon while the var. *pachycereana* can grow on mannitol (man$^+$). In addition, the variety *amethionina* can grow at 27°C in the presence of triterpene glycosides (ttg$^+$) that occur in the stems of cacti of the subtribe Stenocereinae while the variety *pachycereana* is sensitive to these compounds (ttg$^-$) at 27°C. The mannitol variability and triterpene glycoside sensitivity have been shown to be under control of single, independent loci (Starmer *et al.*, 1978b). Strains found in nature are either man$^+$ttg$^-$ or man$^-$ttg$^+$; so far only one strain of mixed phenotype has been isolated. However, combinations of man$^+$ttg$^+$ and man$^-$ttg$^-$ were readily obtained from apparent intervarietal crosses with compatible strains. The disequilibrium of these markers in nature indicates that the gene pools of these varieties are isolated. Intervarietal laboratory crosses using strains isolated from Sonora and Baja California, Mexico, have been reported to yield an average of 35% viable spores (range of viability in different four-spored hybrid asci, 0 to 100%). In cases where conjugation and zygote formation were abundant, the low viability of the spores is apparently due to postzygotic breakdown of meiotic products. Based on the above data, it has been suggested that the *P. amethionina* varieties may be in the final stages of host race formation or early stages of speciation (Starmer *et al.*, 1980). Table III summarizes the DNA homology experiments done with strains of *P. amethionina*. Homology at the intravarietal level approached 100% for *P. amethionina* var. *amethionina, P. amethionina* var. *pachycereana,* and *P. amethionina* strains collected in

Australia, with two notable exceptions that will be discussed below.   The degree of homology at the intervarietal level, however, was more variable.   The varieties of *P. amethionina* var. *amethionina* and *P. amethionina* var. *pachycereana* provided a mean DNA-DNA complementarity value of 64.9 ± 3.2% based on 16 pair-wise comparisons (including reverse experiments).   Strains 76-247B and 76-216 which had been described on the basis of phenotypic properties as *P. amethionina* var. *pachycereana* appear more closely related to *P. amethionina* var. *amethionina* than to the variety *pachycereana*.

The discovery that strains 76-216 and 76-247B (see Starmer *et al.*, 1978b for host plant description) are actually strains that appear rather closely related to *P. amethionina* var. *amethionina* refutes one of the basic notions

Table III.   *Summary of DNA-DNA reassociation values between representative strains of Pichia amethionina*

Organisms	*% relative homology*[a]
P. amethionina var. amethionina (*intravarietal comparisons*)	*100%*
P. amethionina var. pachycereana (*intravarietal comparisons*)	*100%*
P. amethionina strains isolated in Australia from opuntia cacti (*intragroup comparisons*)	*100%*
P. amethionina var. amethionina vs. P. amethionina var. pachycereana strains 76-247B and 76-216 (from columnar cacti in Baja California)[b]	*85%*
P. amethionina var. amethionina vs. P. amethionina strains isolated in Australia	*87%*
P. amethionina var. amethionina vs. P. amethionina var. pachycereana (intervarietal comparisons)	*65%*
P. amethionina var. pachycereana vs. P. amethionina var. pachycereana strains 76-247B and 76-216	*68%*
P. amethionina var. pachycereana vs. P. amethionina strains isolated in Australia	*64%*

[a]*values represent many pair-wise comparisons.*

[b]*for the taxonomic designation of the cacti from which these two strains were isolated, see Starmer et al. 1978b.*

set forth concerning the genetic distance among these
varieties. As mentioned earlier, Starmer *et al.* (1980)
reported that intervarietal crosses of *P. amethionina*
produced spores of which approximately 35% were viable. This
conclusion was based on crosses of a large number of
potentially compatible strains of the two varieties. Many of
the combinations did not produce any viable spores (W.T.
Starmer, personal communication), whereas others produced
spores showing between 50 and 100% viability. The value of
35% viability was taken as an average viability among all
combinations of yeast strains studied. This conclusion now
proves to be inaccurate because all the intervarietal mixes
that produced high proportions of viable spores involved
mixing of 76-247B and 76-216 which were considered to
represent the var. *pachycereana*. In reality, all positive
crosses, on which spore viability was based, were more nearly
intravarietal in nature. Apparently, matings between true
strains of *P. amethionina* var. *amethionina* and the var.
pachycereana do not produce viable spores. If one considers
lack of fertility as criterion for species recognition then
these varieties should be given species status. On the other
hand, DNA complementarity of 65% between the varieties
suggest a rather close relationship or as Starmer *et al.*
(1980) suggested "the beginning of the speciation stage."
Unfortunately, the phenotypes of strains 76-247B (man$^+$ ttg$^+$)
and 76-216 (man$^+$ ttg$^-$) show that a physiologic phenotype
delimiting all strains of *P. amethionina* is not available.

The Australian isolates of *P. amethionina* from *Opuntia*
cacti are more similar to the variety *amethionina* (ca. 87%
homology) than to the variety *pachycereana* (ca. 64%
homology). These strains are man$^+$ (latent) and ttg$^-$
indicating that by current criteria they belong to the
variety *pachycereana*.

Sexual compatibilities among *P. amethionina* var.
amethionina strains isolated in Australia and Mexico have not
been conducted because the Australian strains are homothallic
(W.T. Starmer, personal communication). The base sequence
homology between the Australian strains and the type strain
of *P. amethionina* var. *amethionina* indicates divergence and
the possibility of giving the Australian strains varietal
status. Such status could possibly be rationalized on the
basis of host plant utilization, geographic distribution
(even though relatively recently introduced into Australia)
and the phenotype as illustrated by DNA complementarity with
the type cultures of the other two varieties of *P.
amethionina*.

The relationship of the Australian strains with strains 76-247B and 27-216 (*P. amethionina* var. *amethionina*) has not been investigated as yet. Additional data concerning the status of the Australian strains could be provided by a systematic study of the yeasts associated with *Opuntia* species in the New World, presumably the source of the Australian opuntias.

C. *Pichia opuntiae*

Pichia opuntiae has been isolated from members of the subtribe Pachycereinae (tribe Pachycereeae) and *Opuntia* species (tribe Opuntieae) (Starmer *et al.*, 1979). These strains can be delineated from *P. heedii, P. amethionina,* and *P. cactophila* by their utilization of salicin and/or cellobiose. Two varieties have been described: *P. opuntiae* var. *opuntiae* occurs in *Opuntia* species, whereas *P. opuntiae* var. *thermotolerans* occurs in senita, cardon, and *Pachycereus pecten-aboriginum* (hecho). Phenotypic differentiation of the varieties is based on maximum growth temperature, citrate utilization, rate of cellobiose and mannitol utilization, and reaction to triterpene glycosides (Table I). *P. opuntiae* var. *thermotolerans* occurs in columnar cacti in the Sonoran Desert as a heterothallic, haploid, non-sporulating species. Intravarietal matings produced very few asci. *P. opuntiae* var. *opuntiae* was isolated in Australia and intravarietal mating tests resulted in copious numbers of zygotes and a high percentage of viable spores (strains isolated from nature are usually heterothallic haploid yeasts). Subsequently, strains of each variety were mixed and a mating reaction was observed. Sparse conjugation resulted with an occasional zygote formed. Mating types of the var. *thermotolerans* were designated h^+ or h^- on the basis of their reaction with the mating types of the variety *opuntiae*. The h^+ and h^- strains of the var. *thermotolerans* were then remixed and observed extensively. Spores were observed but only very rarely. Apparently there are factors which have lowered the sexual capability of the variety *thermotolerans* (Starmer *et al.*, 1979). The intervarietal crosses yield few zygotes indicating prezygotic breakdown of fused pairs as the source of incompatibility. The ability of the var. *thermotolerans* to grow at 37°C may have led to its ability to live in the Sonoran Desert. Possibly, the var. *thermotolerans* entered the Sonoran Desert from *Opuntia* cacti. Australian *Opuntia* species are not endemic but have been introduced some 150-200 years ago from New World stock.

The two varieties of *Pichia opuntiae* could possibly represent the final stages in the speciation process (Starmer et al., 1980). Experiments comparing the base sequence similarity of *P. opuntiae* var. *opuntiae* and *P. opuntiae* var. *thermotolerans* are summarized in Table IV. *P. opuntiae* var. *opuntiae* and *P. opuntiae* var. *thermotolerans* are well defined taxa as indicated by the high intravarietal homology values among strains tested. The mean *P. opuntiae* intervarietal relative homology was 28.0 ± 3.1% based on 5 pair-wise comparisons. Only the type cultures of the varieties were used as reference DNA samples in these experiments. The low degree of intervarietal DNA homology in conjunction with earlier data, i.e., lack of interfertility suggests species status for each of the varieties of *P. opuntiae*.

D. *Pichia cactophila*

Pichia cactophila is a host plant generalist, i.e., it has been found in association with all of the cactus species studied thus far. This species is physiologically homogeneous and therefore has been thought to represent a widespread species. Most strains including the type culture are homothallic and form 2 ascospores per ascus while a few strains have been shown to produce 4 spores and these were heterothallic. Table V summarizes the relative homologies among strains of *P. cactophila*.

The type culture appears to be atypical of the species based on these experiments. Homology values greater than 88% between the reference DNA from the type strain and DNA from

Table IV. *Summary of DNA-DNA reassociation values between representative strains of Pichia opuntiae*

Organisms	% relative homology[a]
P. opuntiae var. opuntiae (intravarietal comparisons)	>93%
P. opuntiae var. thermotolerans (intravarietal comparisons)	100%
P. opuntiae var. opuntiae vs. P. opuntiae var. thermotolerans	28%

[a]*values represent many pair-wise comparisons*

other strains were never observed.   Included in these
experiments were *P. cactophila* strains isolated from a
variety of host plants from Baja California, Mexico; Sonora,
Mexico; Australia; Hawaii; Spain; and Venezuela.   The mean
DNA homology value of the above yeasts with the type culture
used as reference was 84.6 ± 1.8% based on 11 pair-wise
comparisons (reverse experiments are presented later).   The
type culture represents an isolate from Rancho San Martin
area of Baja California Sur, Mexico, and there are no other
strains from this area with which to compare this strain.

It is also apparent from the data in Table V that another
group of yeasts has diverged from the type culture of *P.
cactophila*.   These 6 strains show 34.0 ± 2.7% DNA
complementarity with the type strain based on 8
comparisons.   Another homology experiment showed these
strains to be nearly 100% homologous among themselves except
for two strains isolated in Arizona that showed 76% homology
with strains from Baja California and southern Sonora.
Available data indicate that these strains are heterothallic
diploids in nature and produce 4-spored asci (data not
available on all strains), while the type culture and similar
strains (85% homologous) are homothallic diploids and produce
2-spored asci.   Speciation by polyploidy is well known
(White, 1978) but at present we have no data to conclusively
show that this applies to groups within *P. cactophila*.   We
are left with the problem, however, that there are at present

Table V.   *Summary of DNA-DNA reassociation values between representative
strains of P. cactophila*

Organisms	% relative homology[a]
P. cactophila (type strain) vs. P. cactophila strains from Baja California, Mexico; Sonora, Mexico; Spain; Venezuela; Hawaii; and Australia	85%
P. cactophila (type strain) vs. P. cactophila strains producing 4 spores	34% (6 strains)
P. cactophila strains (excluding type culture) intravarietal comparisons	69[b]-100%
P. cactophila strains producing 4 spores among themselves	76-100%[c]

[a]*values represent many pair-wise comparisons*

[b]*the single strain showing 69% homology was isolated in Venezuela (see text)*

[c]*see text*

no known physiological parameters for separating these
strains from *P. cactophila*. Expanding the physiological
parameters used for yeast identification might ultimately
separate such yeasts by means of a simple test.

Experiments examining the similarity of the strains which
were earlier shown to be approximately 85% homologous with
the type culture were conducted. The high degree of homology
among these strains indicates their similarity. A single
strain from Venezuela, showed only 69.2% relative homology
with strains from Baja California, indicating a divergence
possibly due to the evolution of host plants. Included in
these experiments was the reverse experiment involving the
type culture of *P. cactophila* i.e., labeled DNA from one of
the other strains of *P. cactophila*, with unlabeled driver DNA
isolated from the type strain. The reverse comparison, based
on two independent experiments, showed homology values
averaging 96.5%. This type of phenomenon has not been
observed before. This observation can be explained if DNA
isolated from the type culture contains DNA sequences not
present in alternate strains but all the DNA sequences of the
other strains would be present in the type strain. Such DNA
could be independent of or integrated into the nuclear
genome. We have not observed any indication of extra-
chromosomal, extra-mitochondrial DNA in CsCl gradients of DNA
isolated from the type culture but these observations are not
conclusive evidence for the presence or absence of this type
of nucleic acid (2 μ circular DNA has about the same mol% G+C
as the nuclear DNA in *Saccharomyces cerevisiae*). Assuming
that the non-complementary sequence(s) in the DNA isolated
from the type strain are associated with the nuclear DNA, one
must wonder about their nature and origin. They could be
plasmid DNA or viral DNA incorporated into the genome. They
may also be the product of sequence or chromosome
duplication, but this would require a very rapid divergence
of one copy of such a duplication in comparison with the
overall DNA. This seems unlikely in light of inter-strain
similarities of the major portion of the genome. Another
possibility that could be envisioned is a recombinational
event with another yeast species and subsequent degradation
of all foreign DNA with the exception of possibly one small
chromosome. An alternate explanation of the data would be
that all the strains other than the type strain 76-243A had
lost a portion of their genetic material. We have many
strains about 85% homologous with 76-243A but no other
isolates demonstrating the properties of the type strain of

P. cactophila. None of these hypotheses have been explored further but will hopefully stimulate interest in such problems in yeasts at a future date.

In summary, the characterization and the relative intravarietal divergence among strains of *P. heedii, P. amethionina,* and *P. opuntiae* proposed by Starmer *et al.* (1979 and 1980) are supported by data gathered in this study. More specifically, the metabolic types of *P. heedii* have greater than 95% DNA complementarity and are closely related; the two varieties of *P. amethionina* show about 65% homology in their DNAs and are more distantly related; they may be regarded as separate species based on lack of interfertility. The DNAs from the two varieties of *P. opuntiae* have about 28% homology and these varieties are the most distantly related and almost certainly represent separate species (see Fig. 1). Relationships among these three species could not be recognized by the DNA-DNA homology technique. In addition, strains described as *P. cactophila* were found to represent two groups of yeasts, *P. cactophila* and a group showing about 34% homology with *P. cactophila.* Some details of the evolutionary framework of Starmer *et al.* (1980) are contrary to results presented here but in general provided a very sound foundation for this study.

REFERENCES

Anderson, R.S., and Ordal, E.J. (1972). *J. Bact. 109,* 696.
Gibson, A.C., and Horak, K.E. (1978). *Annals Missouri Bot. Gard. 65,* 999.
Holzschu, D.L., Presley, H.L., Miranda, M., and Phaff, H.J. (1979). *J. Clin. Microbial. 10,* 202.
Kreger-van Rij, N.J.W. (1970). *In* "The Yeasts-a Taxonomic Study" (J. Lodder, ed.), p. 455. North-Holland Publ. Co., Amsterdam.
Phaff, H.J., Starmer, W.T., Miranda, M., and Miller, M.W. (1978). *Int. J. Syst. Bacteriol. 28,* 326.
Price, C.W., Fuson, G., and Phaff, H.J. (1978). *Microbiol. Rev. 42,* 161.
Starmer, W.T., Phaff, H.J., Miranda, M., and Miller, M.W. (1978a). *Int. J. Syst. Bacteriol. 28,* 318.
Starmer, W.T., Phaff, H.J., Miranda, M., and Miller, M.W. (1978b). *Int. J. Syst. Bacteriol. 28,* 433.

Starmer, W.T., Phaff, H.J., Miranda, M., Miller, M.W., and
  Barker, J.S.F. (1979). *Int. J. Syst. Bacteriol. 29*,
  159.
Starmer, W.T., Kircher, H.W., and Phaff, H.J. (1980).
  Evolution 34, 137.
Starmer, W.T., Phaff, H.J., Miranda, M., Miller, M.W., and
  Heed, W.B. (1982). *Evol. Biol. 14*, 269.
van der Walt, J.P. (1970). *In* "The Yeasts-a Taxonomic Study"
  (J. Lodder, ed.), p. 34. North-Holland Publ. Co.,
  Amsterdam.
White, M.J.D. (1978). "Modes of Speciation". W.H. Freeman
  & Co., San Francisco.
Wilson, A.C., Carlson, S.S., and White, T.J. (1977). *A.
  Rev. Biochem. 46*, 573.

10
Chemical Composition of Cacti and Its Relationship to Sonoran Desert *Drosophila*

Henry W. Kircher

Department of Nutrition and Food Science
University of Arizona
Tucson Arizona

Host plants are major factors in the ecology of phyto-
phagous insects. Newly hatched larvae live in or on particular portions of their plants and obtain nutrients from the
plant tissues, and, for *Drosophila*, also from microorganisms
associated with the decaying tissues. Fraenkel (1959) pointed out that the photosynthetic portions of all plants contain
proteins, carbohydrates, B vitamins, RNA, sterols, polyunsat-
urated fatty acids, minerals and choline, and should be
equally nutritious to insects. The many mono- and oligopha-
gies that do exist are therefore the result of the ubiquitous
non-nutritive secondary products synthesized by plants to
avoid predation. Although this view has been supported and
disputed, various species of *Drosophila* in the Sonoran Desert
use different cacti as feeding and breeding sites (Fellows
and Heed, 1972) and the chemical compositions of these plants
are strikingly different. In this chapter, the chemistry of
seven cacti is first summarized, then compared, and finally
related to *Drosophila* ecology.

I Composition of Cacti

Extraction of the outer, non-woody tissues of fresh cacti
with methanol followed by chloroform and subsequent partition
of the combined extracts between ether and water gave the 3
fractions shown in Table I.

ECOLOGICAL GENETICS AND EVOLUTION
ISBN 0 12 078820 9

TABLE I.   *Composition of Fresh Cactus*

Cactus[a]	Percent H_2O	Percent Insoluble residue	Lipids	MeOH-H_2O soluble	ALK.[b]	T.G.
A. *Saguaro*	87–88	77	2.5	21	+	–
B. *Senita*	81	71	6–7	25	+++	–
C. *Organ pipe*	77–80	61	11	28	–	+++
D. *Agria*	80	57	6.5	36	–	+++
E. *Cina*		49	5.6	46	–	++++
F. *Prickly pear*	85–88	68	2.7	29	(+?)	–
G. *Backebergia*	80	7		13	+	–

[a]*A. Carnegiea gigantea, B. Lophocereus schottii, C. Stenocereus thurberi, D. Stenocereus gummosus, E. Stenocereus alamosensis, F. Opuntia ficus-indica, G. Backebergia militaris (Gibson and Horak, 1978).*

[b]*ALK=alkaloids, T.G.=triterpene glycosides, (+)=present, (-)=absent.*

Unless specified, all values listed in the tables and text are percentages of the original dry weight of the succulent (non-woody) portion of each cactus including its skin. The skin of cacti, which can be up to 1 mm in thickness and contribute a substantial amount to the dry weight of the plant, is non-cellulosic and is probably composed mostly of cutin similar to the skin of agave (Matic, 1956).

Composition of the insoluble fractions from all of the plants is about the same: cellulose (glucose) plus polysaccharides containing galactose, glucose, arabinose and xylose with smaller amounts of mannose, rhamnose and uronic acid (Moyna and di Fabio, 1978; Steelink *et al.*, 1968). This fraction also includes lignin and insoluble inorganic compounds (silica, Ca oxalate) present as well as cutin from the skin of the plant. Chemical compositions of the lipid and water soluble fractions shown in Table I differ between the cacti and will be discussed in more detail. All unreferenced data given in this paper are from unpublished work done in the author's laboratory.

A.   *Carnegiea gigantea*

Saguaro contains 2–3% lipid composed of $\sim$ 1% (as esters) of the typical fatty acids (Fig. 1A) associated with photosynthetic tissue, 1–1.7% of two simple isoquinoline alkaloids (Fig. 1C, Brown *et al.*, 1972) and $\sim$ 0.1% of the typical plant sterols (Fig. 1D).

The methanol-water soluble portion of saguaro (~ 20% of the plant solids) is a poorly defined mixture of substances. Steelink *et al.* (1967,1968) reported ~ 10% dopamine in saguaro as well as small amounts of glucose, fructose, hydroxycinnamic acids and 4-hydroxybenzoic acid glucoside. We were unable to show any characteristic spots on TLC for sterol or triterpene glycosides in this fraction and detected only glucose after acid hydrolysis.

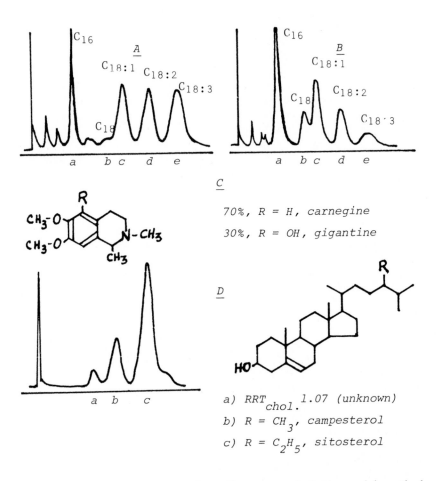

70%, R = H, carnegine

30%, R = OH, gigantine

a) RRT_{chol}. 1.07 (unknown)

b) R = CH_3, campesterol

c) R = C_2H_5, sitosterol

FIGURE 1. GLC separation diagrams of fatty acid methyl esters of A fresh, B decayed saguaro, C saguaro alkaloids, D saguaro sterols. Left, GLC separation diagram; right, sterol structures.

B.   *Lophocereus schottii*

Senita cactus has more lipid than saguaro and contains a number of unusual alkaloids and sterols. The highest concentration of alkaloids is in the green epidermis where they exist as dimers and trimers of the isobutyl substituted isoquinoline, lophocereine (Fig. 2A, Djerassi *et al.*, 1962; Tomita *et al.*, 1963).

Two of the senita sterols, lophenol and schottenol, and the triterpene, lupeol, as well as 0.9% octyl alcohol were characterized by Djerassi *et al.* (1958); the remaining six sterols in senita were identified 20 years later (Campbell and Kircher, 1980) (Fig. 2B). The fatty acids in senita occur as esters and are the ones associated with photosynthetic tissue (Fig. 2C). Concentrations and kinds of alkaloids, non-saponifiables (sterols, terpenes, carotenes) and fatty acids were compared in several tissues of numerous senita cacti over the range of the plants in Sonora, Mexico (Kircher, 1969). The only clines observed were a tendency of shorter chain fatty acids to be more abundant in plants from southern Sonora (Fig. 2D) and more alkaloids in mature stems in the north. The total lipid concentration of the stems was also higher in the outer photosynthetic tissues than in the yellow cortex of the cactus.

Composition of the MeOH-H_2O soluble portion of senita is not known but it probably resembles that of saguaro. Senita cortex tissue also darkens when cut suggesting the presence of dopamine and in one analysis, glucose and fructose were observed in this fraction.

C.   *Stenocereus thurberi*

Organ pipe cactus, previously known as *Lemaireocereus thurberi* (Gibson and Horak, 1978), contains no alkaloids (Djerassi *et al.*, 1953) but rather large quantities of lipids and triterpene glycosides (Table II, Kircher, 1977). The lipids are principally monoesters (C_8, C_{10}, C_{12}, Fig. 3A) of nine neutral pentacyclic triterpenes (Fig. 3B, Jolad and Steelink, 1969; Kircher, 1980) and of five 3β,6α-dihydroxy-sterols (Fig. 3C, Kircher and Bird, 1982). The three components, acids, triterpenes and sterol diols, occur as an approximate 1:2:1 ratio in organ pipe lipids. In addition to these, about 0.07% of the outer tissues is a 1:2:7 mixture of cholesterol, campesterol and sitosterol (Kircher, 1980). The triterpene glycosides, which are the main constituents in the MeOH-H_2O soluble fraction, are composed of three acidic

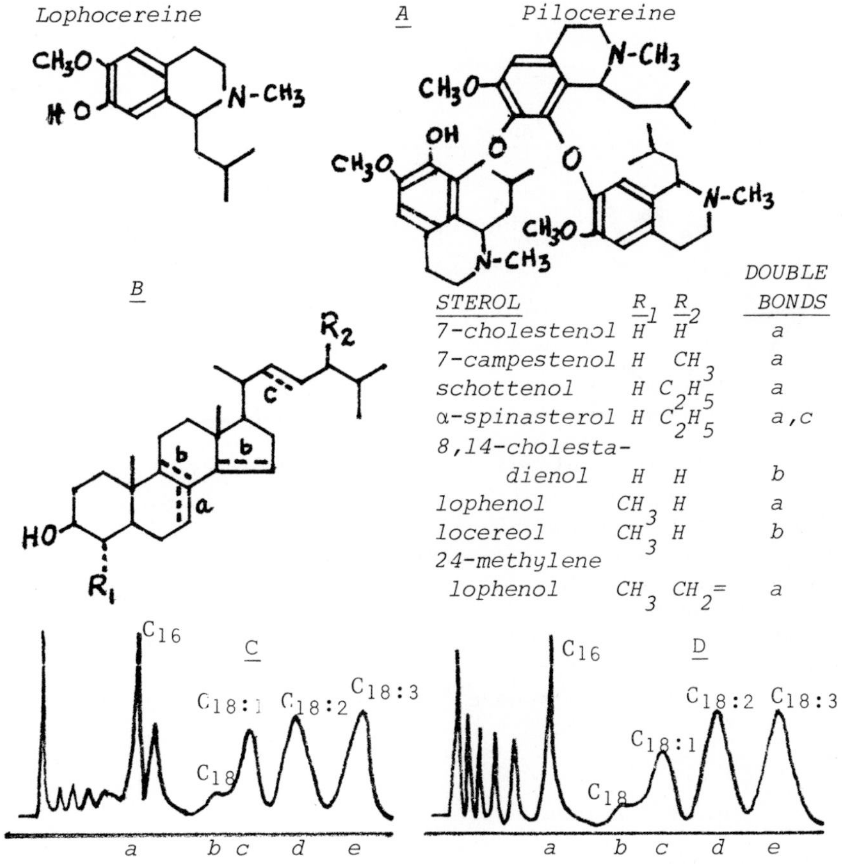

Lophocereine
A
Pilocereine
CH_3O
HO
N-CH_3
OH
O
B
R_2
R_1
HO
a
b
c
STEROL | R_1 | R_2 | DOUBLE BONDS
7-cholestenol H H a
7-campestenol H CH_3 a
schottenol H C_2H_5 a
α-spinasterol H C_2H_5 a,c
8,14-cholesta-dienol H H b
lophenol CH_3 H a
locereol CH_3 H b
24-methylene lophenol CH_3 CH_2= a
C
C_{16}
$C_{18:1}$
$C_{18:2}$
$C_{18:3}$
C_{18}
a b c d e
D
C_{16}
$C_{18:1}$
$C_{18:2}$
$C_{18:3}$
C_{18}
a b c d e

FIGURE 2. Senita lipid components. A. Alkaloids, B. Sterols, Fatty acid methylesters from: C. Northern, D. Southern Sonora.

pentacyclic triterpenes (Fig. 3B) each attached to a tetrasaccharide made from two glucose and two rhamnose residues (Djerassi *et al.*, 1953; Marx *et al.*, 1967; Kircher, 1977).

D. *Stenocereus gummosus*

Agria cactus, formerly known as *Machaerocereus gummosus* (Gibson and Horak, 1978) has not been analyzed as completely as the previous three cacti but the essential components in the lipid and methanol soluble fractions are known. Stems of the plant contain no alkaloids but are very rich in MeOH-H_2O

TABLE II.   *Composition of a Cross Section of a Mature Stem*
of Organ Pipe Cactus (g/100 g dry tissue)

Tissue	Lipids	Triterpene glycosides	Insoluble residue
Skin	2.8	5.0	92.2
Photosynthetic layer	12.2	50.5	37.3
Transition zone	10.0	34.2	55.8
Cortex	8.1	15.0	76.9
Wood	3.6	3.5	92.9
Pith	7.3	12.6	80.1

soluble glycosides of one neutral and two acidic triterpenes
(Fig. 4A, Djerassi *et al.*, 1954, 1955). Examination of the
lipid fraction showed only small amounts of normal phytoste-
rols but relatively large quantities of sterol diols similar
to those in organ pipe and of three pentacyclic triterpenes
(Fig. 4A) monoesterified to medium chain fatty acids (Fig. 4B)
as well as smaller amounts of several unknown compounds. Hydrolysis of the triterpene glycosides gave an approximate 2:1
ratio of glucose: rhamnose and a 4:1 ratio of sugars:
aglycones.

E.   *Stenocereus alamosensis*

Cina cactus, formerly known as *Rathbunia alamosensis* (Gibson and Horak, 1978), is a plant with a narrower stem diameter than agria and grows in both Sonora and Baja California.
It is very rich in triterpene glycosides (Table I) which upon
hydrolysis, gives the same triterpenes that are present in
agria (Fig. 4A). Hydrolysis of the lipids also yielded the
same neutral compounds present in agria and distinctly different from those in organ pipe cactus. Although the sample
of cina examined had more longispinogenin (Fig. 4A) and less
sterol diols than agria, this observation may be a result of
natural variations in the concentration of these compounds
within each species. Also the concentration of gummosogenin
relative to that of machaeric and machaerinic acids (Fig. 4A)
was much higher in cina than in agria triterpene glycosides.
In addition, glucose was the only sugar detected in the cina
derivatives whereas glucose and rhamnose were present in organ pipe and agria triterpene glycosides.

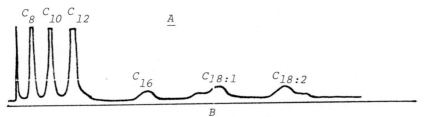

$\underline{A}$

$R_1, R_2 = H; R_3, R_4 = CH_3$ unless otherwise noted

$\underline{B}$

Triterpenes in Lipids

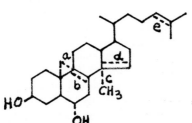

$\underline{Lupenes}$

$\underline{Oleanenes}$

Lupeol (L)
Betulin (L, $R_3 = CH_2OH$)*
Calenduladiol (L, $R_1 = OH$)*
Lupenetriol (L, $R_1 = OH, R_3 = CH_2OH$)
Betulinic aldehyde (L, $R_3 = CHO$)
Methyl betulinate (L, $R_3 = COOCH_3$)
Longispinogenin (O, $R_1 = OH$, $R_3 = CH_2OH$)
Oleanolic aldehyde (O, $R_3 = CHO$)
Methyloleanolate (O, $R_3 = COOCH_3$)

Triterpenes in glycosides

Oleanolic acid (O, $R_3 = COOH$)*
Queretaroic acid (O, $R_3 = COOH$, $R_4 = CH_2OH$)*
Thurberogenin (L, $R_2, R_3 = -O-CO-$, lactone)*

$\underline{C}$

Sterol diols in Lipids

Cyclosterol (a,c)
Macdougallin (b,c)*
Stenocereol (b,c,e)*
Thurberol (b,d)*
Peniocerol (b)*
*Major components

FIGURE 3. *Components of organ pipe lipids and triterpene glycosides: A. GLC separation diagram of fatty acid methyl esters, B. Triterpenes, C. Sterol diols.*

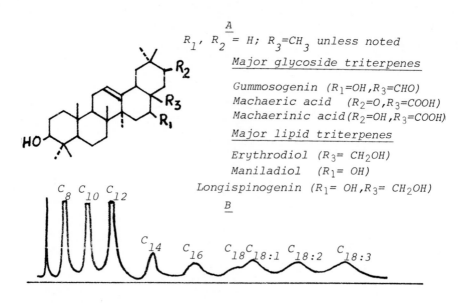

R_1, R_2 = H; $R_3=CH_3$ unless noted

Major glycoside triterpenes

Gummosogenin ($R_1=OH$, $R_3=CHO$)
Machaeric acid   ($R_2=O$, $R_3=COOH$)
Machaerinic acid($R_2=OH$, $R_3=COOH$)

Major lipid triterpenes

Erythrodiol ($R_3= CH_2OH$)
Maniladiol   ($R_1= OH$)
Longispinogenin ($R_1= OH$, $R_3= CH_2OH$)

Figure 4.   Components of agria triterpene glycosides and lipids: A triterpenes, B GLC separation diagram of fatty acid methyl esters.

F.   Opuntia ficus-indica

This species of prickly pear is known worldwide. It is grown as an ornamental in Tucson and is a pest in South Africa. Various species of Opuntia have been analyzed for a number of constituents but a thorough investigation of any of them is lacking. Cruse (1973), in a review of desert plants, listed 0.04% sitosterol and a mixture of C_{12}-C_{18} free fatty acids in O. ficus-indica flowers and glucose, glucose-6-phosphate isomerase, a lysine deficient but otherwise nutritionally acceptable amino acid profile and a homogeneous mucilage containing a 3:3:1:1 ratio of galactose, arabinose, rhamnose and xylose in the cladodes (stems). This mucilage, obtained in 0.3 to 1% yield from the plant was investigated more recently (McGarvie and Parolis, 1981) and shown to be a highly branched molecule. Short chains of galactose and arabinose with arabinose and xylose terminals are attached to long chains of rhamnose and galacturonic acid. In another recent study, 5 of 6 species of platyopuntia, including O. lindheimeri, O. phaeacantha and O. stricta, were shown to contain primary and secondary amine alkaloids but none could be identified by TLC with known compounds (Meyer et al., 1980). The fatty acids in the cladodes of *O. engelmanii* were principally the ones associated with photosynthetic tissue: C_{16}, $C_{18:1}$, $C_{18:2}$, $C_{18:3}$ in a 2:1:1:1 ratio (Pieters, 1972).

In a preliminary analysis of *O. ficus-indica*, young and mature cladodes contained 87.4 and 85.4% water, respectively. Glucose, fructose and smaller amounts of galactose were detected in an aqueous extract of the plant with glucose, galactose and xylose evident after hydrolysis of the insoluble residue (Table I). The 2.7% lipids were hydrolyzed to a 1:1 ratio of fatty acids and non-saponifiables. The former were composed of the usual plant fatty acids (C_{16}, $C_{18:1}$, $C_{18:2}$, $C_{18:3}$ 2:1:3:2) and the latter of sitosterol, sitosteryl glycoside and a small amount of unknown alkaloid.

G. *Backebergia militaris*

Only one sample of this cactus has been analyzed; it was brought to Tucson from Michoacan, Mexico by A.C. Gibson in 1978. The alkaloids were determined to be 3,4-dimethoxy-β-phenylethylamine (0.04%) and heliamine (0.75%) (Fig. 5A, Mata and McLaughlin, 1980). In our study we found the typical plant fatty acids plus what appeared to be lauric acid (C_{12}) (Fig. 5B) after hydrolysis of the lipids and the same sterols that are present in senita cactus (Fig. 5C) in the non-saponifiable fraction. No triterpene glycosides were present in the MeOH-H_2O soluble fraction.

II Comparison of the Cacti

The seven plants discussed here fall into two main groups, those that contain alkaloids, saguaro, senita, prickly pear and *Backebergia,* and those that contain triterpene glycosides, organ pipe, agria and cina. In the first group, saguaro and *Backebergia* contain ∿ 1% of simple alkaloids whereas senita has a much greater quantity (3–10%) of larger, more complicated alkaloids. Saguaro and prickly pear resemble each other in their low lipid and typical phytosterol and fatty acid contents; senita and *Backebergia* both are relatively rich in lipids and unusual Δ^7 and $\Delta^{8,14}$-sterols. Senita also differs from *Backebergia* in having much less lauric acid in its lipids.

Organ pipe, agria and cina all contain large (25–45%) quantities of water soluble triterpene glycosides and lipids

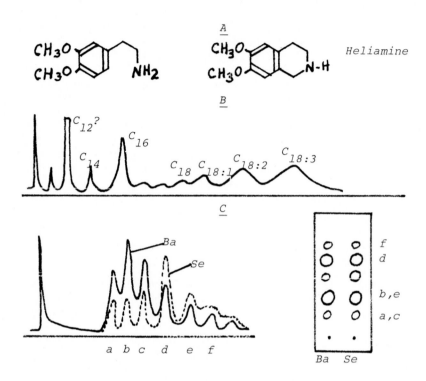

FIGURE 5.   *Backebergia militaris lipids: A.   alkaloids,*
B.   GLC diagram of fatty acid methyl esters, C.   GLC (left)
and TLC on AgNO₃ plates (right) of Backebergia and senita
non-saponifiables: a) 8,14-cholestadienol, b) 7-cholestenol,
c) locereol, d) lophenol, e)schottenol, f) lupeol.

(7-12%).   The triterpenes in these two fractions differ between the plants.   In organ pipe they are mainly lupenes, and
in agria and cina, oleanenes.   In all three cases, the sugar
moiety of the glycosides is a glucose-rhamnose oligosaccha-
ride and the sterol diols appear to be the same as are the
fatty acids.   On the basis of limited evidence, cina and
agria appear to have a higher concentration of glycosides in
their tissues and organ pipe more of the lipids with a relatively greater proportion of medium chain fatty acids esteri-
fied to its neutral triterpenes and sterol diols.
      In summary, although the following comparisons are sub-
jective and may not be ecologically meaningful, the seven
cacti can be listed in an order of *decreasing* chemical complexity: Organ pipe > Cina > Agria > Senita > *Backebergia*
militaris > Saguaro > *Opuntia ficus-indica* with the last

succulent closest in composition to that of a "typical plant."

When saguaro, organ pipe, senita and agria were analyzed before and after decaying (Table III), the principal substances consumed in each case were non-cellulosic carbohydrate polymers and the sugars of the glycosides. Unsaturated fatty acids in saguaro also appeared to be selectively consumed (Fig. 1B) and a portion of the organ pipe lipids were hydrolyzed but were not used as a carbon source by the bacteria, yeast and mold causing the decay. The senita sterols but not the alkaloids could still be identified in an extract of the dry soil under the woody skeleton of a completely decayed senita.

III Cacti and *Drosophila* Ecology

Collections and rearing records of desert adapted *Drosophila* (Fellows and Heed, 1972; Heed, 1978) demonstrated a number of host plant specificities between particular species in this genus and cacti. *D. nigrospiracula* is found only in decaying tissues of saguaro and its close relative, cardón (*Pachycereus pringlei*) and *D. pachea* only in senita. *D. mettleri* appears to be behaviorally restricted to soil breeding, but not necessarily to soil soaked by a particular cactus rot (Fogleman *et al.*,1981). *D. mojavensis* has the broadest distribution. It breeds in organ pipe cactus in Arizona and Sonora where agria is absent but preferentially in agria where both plants exist. It is also found in decaying

TABLE III. Chemical Changes during Cactus Decay[a]

| Cactus | Percent of each fraction lost[b] | | | Percent of total dry wt. lost |
	Insoluble residue	Lipid	H_2O-MeOH soluble	
Saguaro	26	38	48	31
Organ Pipe	27	-18[c]	63	33
Senita	26	-67[c]	48	27
Agria	24	-90[c]	51	29

[a]*Average of two determinations.*
[b]*Refer to Table I for composition of these cacti before decay.*
[c]*Relative gain in weight of this fraction: Hydrolysis of glycosides.*

barrel cacti in the Anza-Borrego Desert in southern California and in *Opuntia* species on the Catalina Islands near Los
Angeles (Heed, pers. comm.). *D. arizonensis* is the principal
species found in decaying cina cactus. No *Drosophila* has
been found as yet associated with *Backebergia militaris,* but
that is because the cactus grows in southwestern Mexico and
its decaying tissues have not been adequately examined. *D.
hamatofila, D. longicornis and D. arizonensis* inhabit decaying *Opuntia* cacti in the Sonoran Desert whereas *D. buzzatii*
and *D. aldrichi* use these plants as breeding sites in Australia. The reasons why these specificities exist probably go
beyond chemical and physical differences between the decaying
cacti to include temperature, humidity, response to volatiles,
predators and microorganisms, but in this chapter chemical
rationales will be the only considerations discussed.

A.   *D. pachea*

The restriction of *D. pachea* to senita cactus is a result
of the strict Δ^7-sterol requirement of the fly (Heed and Kircher, 1965) and its tolerance of the toxic senita alkaloids
(Kircher *et al.*, 1967). In our original study, only 7-stig-
mastenol (schottenol) and lophenol (4α-methyl-7-cholestenol)
were known to be present in the cactus and the synthetic
schottenol we prepared at the time was later shown to be con-
taminated with 40% 7-campestenol (Fig. 2B). Subsequently,
experiments with pure schottenol demonstrated that this sterol does not allow survival of *D. pachea* larvae past the second instar and that 7-cholestenol and 7-campestenol, which
are also present in senita (Campbell and Kircher, 1980), are
the sterols used by the insect (Table IV).

TABLE IV. Effect of Δ^7-Sterols on D. pachea

Sterol in the diet	Side chain R_1	Side chain R_2	No. of adults F_1	No. of adults F_2
None	–		1200	0
7-cholestenol	H	H	936	1750
7-campestenol	CH_3	H	734	1119
7-ergostenol	H	CH_3	295	0
7-stigmastenol (schottenol)	C_2H_5	H	367	0

B. *D. nigrospiracula*

The close association of *D. nigrospiracula* with saguaro and cardón can be explained on a chemical basis only from the standpoint that this species is excluded from other potential hosts. As far as is known there is nothing inherently beneficial in these cacti for *D. nigrospiracula* except the absence of potentially toxic compounds. This fly cannot tolerate senita alkaloids (Kircher *et al.*, 1967) or organ pipe (and presumably agria) medium chain fatty acids and sterol diols (Table V). It should be able to breed in *Opuntia* based on known prickly pear cactus chemistry but perhaps this plant does not rot adequately for *D. nigrospiracula* or the slime producing, water soluble polysaccharides of *Opuntia* interfere with larval development.

TABLE V.   *Effect of Organ Pipe Cactus Constituents on D.*
           nigrospiracula and D. mojavensis

Species	Numbers of F_1 emerging from each medium Percent organ pipe fatty acids					
	none	0.5	0.75	1.0	1.5	2.0
D. nigrospiracula	188	37	74	0	0	0
D. mojavensis	692	852	761	755	500	0

	Percent organ pipe sterol diols				
	none	1.5	3.5	5.0	10.0
D. nigrospiracula	335	46	19	0	0
D. mojavensis	>500	—	∿500	792	755

C. *D. mettleri*

This species breeds mainly in soil soaked by decaying saguaro and cardón although it has been isolated once from soil under a decaying organ pipe (Fogleman *et al.*, 1981). When first instar larvae were placed on soils artificially soaked with juices from rotting *Opuntia*, organ pipe and agria in the laboratory, *D. mettleri* successfully eclosed (Fogleman, pers. comm.). It is also able to breed in senita soaked soil and is resistant to the toxicity of senita alkaloids (Fogleman *et al.*, 1982) and therefore has the greatest number of potential niches of all the desert *Drosophila*. The absence or rarity of suitably soaked soil under any of the decaying cacti except saguaro and cardón, however, effectively limits this fly's larval habitat. We do not yet know if *D. mettleri* is able to use the senita Δ^7 and $\Delta^{8,14}$-sterols; in laboratory tests with a few cholesterol analogs, it produced the most progeny with sitosterol (Fig. 1D) in its medium and appeared to be the most sensitive of four species tested (with *D. melanogaster, D. pseudoobscura, D. mojavensis)* to subtle changes in the sterol molecule.

D.   *D. mojavensis*

This species is prevented from breeding in senita by toxic alkaloids (Kircher *et al.*, 1967) and in saguaro by *D. nigrospiracula* (Fellows and Heed, 1972; Mangan, 1982). The lipids and triterpene glycosides of organ pipe, when added to standard medium in concentrations equivalent to those in the cactus, do not affect reproduction of *D. mojavensis* and it is even able to produce F_1, albeit stunted in size, on a medium containing as much as 60% of its total dry weight of triterpene glycosides. In nature, *D. mojavensis* emerging from agria are smaller than those from organ pipe (Heed, pers. comm.); this may be a result of the higher glycoside concentration in agria.

As organ pipe rots, the lipids and triterpene glycosides are hydrolyzed (Table III) with the release of sterol and triterpene diols and free fatty and triterpene acids. The diols did not affect *D. mojavensis* larval development at any of the levels tested and the free medium chain fatty acids were toxic only at levels of 2% or greater (Table V), a level rarely reached in natural organ pipe rots.

E.   *Backebergia militaris*

This cactus, even though it does not occur in the Sonoran desert, is included here because of its interesting chemistry. In its low concentrations and types of alkaloids, it resembles saguaro, cardón and *Opuntia*. It has no triterpene glycosides but its sterols appear to be identical to those in senita (Fig. 5C) and it may have a relatively high concentration of lauric acid in its lipids. It may be a better substrate than senita for *D. pachea* and better than organ pipe or agria for *D. mojavensis*. If *D. nigrospircula* and *D. mettleri* are able to use its sterols, *Backebergia* should also be suitable for them. All this remains to be tested as well as the determination of what species of *Drosophila*, if any, use the decaying tissues of this cactus.

F. *Opuntia ficus-indica*

A number of non-endemic Sonoran Desert *Drosophila* have been reared from decaying *Opuntia* pads. *O. ficus-indica* decays with the development of a viscous, slimy liquid in the rot which may be a product of bacterial breakdown of the cell walls without hydrolysis of the water soluble arabogalactan in this plant. Similar polysaccharides are exuded by wounded trees and cyclindriopuntias (cherry gum, gum acacia, cholla gum) and may immobilize bacteria to prevent infection of these plants. Since the *O. ficus-indica* polysaccharide is already present in fresh tissue and appears to be maintained during early stages of the rotting process, it may inhibit certain *Drosophila* larvae by coating them with slime.

G. *L'envoie*

The cactus-microorganism-*Drosophila* relationships are complex and not all of them will be explicable by chemical rationales. Nevertheless, since the initiation and maintenance of the chain ultimately depend on the host cacti, a study of cactus chemical composition before and during decay and its correlation with their associated microorganisms and species of *Drosophila* will continue to be a worthwhile endeavor. This habitat variety within a restricted locale for species within a single genus makes the system particularly attractive in the field of insect-host plant interactions.

REFERENCES

Brown, S.D., Hodgkins, J.E., Massingill, J.L., Jr., and Reinecke, M.G. (1972). *J. Org. Chem. 37*, 1825.

Campbell, C.E., and Kircher, H.W. (1980). *Photochemistry 19*, 2777.

Cruse, R.A. (1973). *Econ. Bot. 27*, 210.

Djerassi, C., and Lippman, A.E. (1955). *J. Am. Chem. Soc. 77*, 1825.

Djerassi, C., Geller, L.E., and Lemin, A.J. (1953). *J. Am. Chem. Soc. 75*, 2254.

Djerassi, C., Geller, L.E., and Lemin, A.J. (1954). *J. Am. Chem. Soc. 76*, 4089.

Djerassi, C., Krakower, G.W., Lemin, A.J., Liu, H.H., Mills, J.S., and Villotti, R. (1958). *J. Am. Chem. Soc. 80*, 6284.

Djerassi, C., Brewer, H.W., Clark, C., and Durham, L.J. (1962). *J. Am. Chem. Soc. 84*, 3210.

Fellows, D.P., and Heed, W.B. (1972). *Ecology 53*, 850.

Fogleman, J.C., Hackbarth, K., and Heed, W.B. (1981). *Am. Nat. 118*, 541.

Fogleman, J.C., Heed, W.B., and Kircher, H.W. (1982). *Comp. Biochem. Physiol.* (in press).

Fraenkel, G. (1959). *Science 129*, 1466.

Gibson, A.C., and Horak, K.E. (1978). *Ann. Missouri Bot. Gard. 65*, 999.

Heed, W.B. (1978). *In* "Ecological Genetics: The Interface" (P.F. Brussard, ed.), p. 109. Springer Verlag, New York.

Heed, W.B., and Kircher, H.W. (1965). *Science 149*, 758.

Jolad, S.D., and Steelink, C. (1969). *J. Org. Chem. 34*, 1367.

Kircher, H.W. (1969). *Phytochemistry 8*, 1481.

Kircher, H.W. (1977). *Phytochemistry 16*, 1078.

Kircher, H.W. (1980). *Phytochemistry 19*, 2707.

Kircher, H.W., and Bird, H.L., Jr. (1982). *Phytochemistry* (in press).

Kircher, H.W., Heed, W.B., Russell, J.S., and Grove, J. (1967). *J. Insect Physiol. 13*, 1869.

Mangan, R. (1982). Chapter 17, this Volume.

Marx, M., Leclerq, J., Tursch, B., and Djerassi, C. (1967). *J. Org. Chem. 32*, 3150.

Mata, R., and McLaughlin, J.L. (1980). *J. Pharm. Sci. 69*, 94.

Matic, M. (1956). *Biochem. J. 63*, 168.

McGarvie, D., and Parolis, H. (1981). *Carb. Res. 88*, 305.

Meyer, B.N., Mohamed, Y.A.H., and McLaughlin, J.L. (1980). *Phytochemistry 19*, 719.

Moyna, P., and di Fabio, J.L. (1978). *Planta Med. 34*, 207.

Pieters, E.P. (1972). *Phytochemistry 11*, 2623.

Steelink, C., Yeung, M., and Caldwell, R.L. (1967). *Phytochemistry 6*, 1435.

Steelink, C., Riser, E., and Onore, M.J. (1968). *Phytochemistry 7*, 1673.

Tomita, M., Kikuchi, T., Bessho, K., and Inubushi, Y. (1963). *Tet. Lett.*, 127.

11
Associations and Interactions Among Yeasts, *Drosophila* and Their Habitats[1]

William T. Starmer

Department of Biology
Syracuse University
Syracuse, New York

I Introduction

Most yeasts are saprophytic and serve to transform decaying organic matter into food items for other organisms in
the food web. The manner in which yeasts obtain energy and
the consequences of their metabolism are thus important to
the organisms that utilize them as a source of nutrition.
In general the physiological capacity of yeasts determines
to a large extent the distribution of yeast species over
space and through time (Phaff and Starmer, 1980). The interaction of host (e.g. decaying plant tissue) chemistry and
the physiological abilities of the yeasts presumably determines the ecological specificity of yeast species (Starmer,
1981a, 1981b). Rapid exploitation of free sugars is a common characteristic of most yeast species and this property
has led to the general observation that yeasts are found
among the first microorganisms in rotting plant tissue.
Coupled with this colonization character is the general re-
liance of yeasts on insects for directed dispersal to new
habitats (Gilbert, 1980). Therefore the specificity of the
insects also plays a role in the specificity of the yeasts
transported by the insects. Since the plant chemistry determines to some extent the types of saprophytic insects which

[1]Supported by NSF grant DEB 78-22041 to W. T. Starmer

ECOLOGICAL GENETICS AND EVOLUTION
ISBN 0 12 078820 9

utilize the yeasts within the decaying tissue, it might be
expected that the yeasts and their vectors have coevolved
with respect to the chemicals found in their host plants.
This paper outlines and documents a probable mode of coadaptation experienced by lipolytic yeasts and *Drosophila mojav-*
ensis, both of which utilize the decaying stems of *Steno-*
cereus thurberi (organpipe cactus).

Drosophila mojavensis breeds and feeds on a variety of
columnar cacti found in the Southwestern United States and
Northwestern Mexico (Heed, 1978). In Baja California, Mexico, *D. mojavensis* utilizes agria cactus (*Stenocereus*
gummosus), while in Sonora, Mexico and Southern Arizona,
USA the fly utilizes organpipe cactus. Studies on the larval feeding behavior of *D. mojavensis* have shown that larvae
selectively feed on the yeast *Pichia cactophila*. This is
true for larvae in their natural habitat and for larvae
given choices of yeasts in the laboratory (Fogleman *et al.,*
1981). However, *D. mojavensis* does include other yeast
species in its diet.

Nineteen species and varieties of yeasts have been recovered from cactus necroses utilized by *D. mojavensis* (see
Table 1 in Fogleman *et al., 1981*). Five of these species
are found in relatively high frequency. Only two of these
species (*Candida ingens* and *Pichia mexicana*) produce extracellular lipases. This characteristic is presumed to be
related to the lipid composition of the cacti in which these
yeasts occur most frequently (Miranda *et al.,* 1982).

Organpipe cactus contains unusual lipids which have been
placed in two chemical groups (Kircher, 1980): (1) medium
chain fatty acid esters of sterol diols; (2) medium chain
fatty acid esters of triterpene diols. These major components of the lipid fraction account for 10-15% of the tissue
(dry weight basis). The medium chain fatty acids account for
10% of the lipids and are made up of caproic (C6), caprylic
(C8), capric (C10) and lauric (C12) acids.

This paper will specifically address the question, do the
lipolytic yeasts and lipids of organpipe cactus interact to
affect the fitness characteristics of *D. mojavensis?*

II   Materials and Methods

A population of *D. mojavensis* collected during December,
1979 near San Telmo, Baja California, Mexico was utilized in
all experiments. This strain (A763) was rendered axenic by

dechorionation of eggs with clorox and was maintained axen-
ically on sterile food for the duration of the experiments.
First instar larvae (less than 12 hours old) were transferred
from the axenic culture to test media. Depending on the
experiment 20, 25 or 30 larvae were transferred to test vials.
Adults were removed the day of eclosion, placed on banana
food, aged 24 hours, and then stored in a freezer until measurements were made. In most cases adults were sexed, counted
and thorax length measured with an ocular micrometer (1
ocular unit = 0.039 mm). Percent emergence per vial was
transformed by the arcsin square root transformation for
analysis. The following microorganisms were employed in the
experiments: *Erwinia carnegieana* (a bacterium known to cause
cacti to rot, Lightle *et al.*, 1942); the yeasts (1) *Pichia
cactophila*, (2) *Candida sonorensis* (3) *Pichia amethionina*
var. *amethionina* (4) *Candida ingens* and (5) *Pichia mexicana*.

These yeasts were tested for their ability to (1) use
organpipe lipids as a source of carbon by inoculating them
on YNB (yeast nitrogen base, Difco) supplemented with 2%
agar and 3% organpipe lipids, (2) tolerate organpipe lipids
by inoculating them on a complete medium, YM agar (yeast ex-
tract-malt extract agar, Difco) supplemented with 3% organpipe lipids, (3) utilize caproic and caprylic acid as a carbon source at various pH levels by inoculating them on YNB
agar supplemented with 0.05 or 0.1% of the sodium salt of the
acid and (4) tolerate these fatty acids by inoculating them
on YNB agar supplemented with 0.5% fructose as the carbon
source and 0.05 or 0.1% of the sodium salt of the acid at
various pH levels. The effect of pH was ascertained by testing without the fatty acid. All tests were conducted at 25°C.
Concentrations of lipids and fatty acids approximate the
levels known to occur or to be expected in organpipe cactus.

A. Experiment with Organpipe Cactus Tissue (Experiment A)

Ten grams of fresh (frozen) tissue of organpipe was
placed onto 10 grams of sand in 8 dram shell vials, plugged
with cotton and autoclaved. Three vials were untreated
(control), while 18 vials were inoculated with *E. carnegieana* and incubated at 27° for 48 hours. Three of these 18
vials were not treated further (bacterial treatment) while
the remaining 15 vials were inoculated in sets of three with
each of the five yeasts. All 21 vials were then seeded with
25 larvae of the axenic strain of *D. mojavensis*.

B. *Experiment with Organpipe Lipids (Experiment B)*

The lipid fraction of organpipe was blended with sterile banana food at concentrations of 0, 1.5 (approximate natural concentration) and 3 percent (Lipid treatment). Fructose was added to this medium at 0.5% (w/v) to insure a carbon source was present for yeast growth. Two vials of each concentration were either uninoculated or inoculated with yeast (5 species). All 36 vials were then seeded with 30 larvae of the axenic strain of *D. mojavensis*.

C. *Experiment with Medium Chain Fatty Acid (Experiment C)*

Caprylic acid (representative fatty acid of the lipid component) was added to sterile banana food at levels of 0, 1x, 2x, 3x and 4x such that 1x = 1.5 mMoles. Addition was accomplished by adding a 1 ml solution of ethanol and appropriate amount of acid to sterilized warm banana medium just before it was mixed with sterile agar and distributed to vials. Fructose was also added at 0.5% (w/v) to insure a carbon source was present for yeast growth. Three vials of each fatty acid level were retained as controls while three vials of each level were inoculated with either *C. ingens* or *P. amethionina*. These 45 vials were then seeded with 20 larvae of the axenic strain of *D. mojavensis*.

III   Results

Table I lists the five common cactus yeasts associated with organpipe cactus. This table is an updated (as of December, 1981) and abbreviated version of Table I presented by Fogleman *et al.* (1981). It is clear that *P. cactophila* and *C. sonorensis* are the dominant yeasts in the cactus stems utilized by *D. mojavensis*. It is also apparent that the lipolytic yeasts *C. ingens* and *P. mexicana* are most frequent in organpipe cactus as compared to other cacti.

TABLE I.   *The Frequency (no. of Isolates/no. of Plants
Sampled) of the Most Common Yeasts Associated
with Cacti Utilized by D. mojavensis*

CACTUS	*Stenocereus gummosus*	*Stenocereus thurberi*	*Myrtillocactus cochal*	*Ferocactus acanthodes*
Number of Plants	116	46	15	13
YEAST				
Pichia cactophila	0.75	0.54	0.73	1.00
Candida sonorensis	0.46	0.44	0.80	0.23
Pichia amethionina				
var. amethionina	0.24	0.09	0.07	0.08
Candida ingens	0.10	0.33	0	0.08
Pichia mexicana	0.05	0.22	0.07	0

Table II lists the results of tests designed to determine the yeast's ability to utilize and tolerate the lipids and medium chain fatty acids found in organpipe cactus. This table shows that all yeasts can utilize the lipid fraction to some degree but that *C. ingens* grows best. Furthermore, all yeasts can tolerate the lipids at their natural concentration.

Tests conducted with the sodium salts of caproic (C6) and caprylic (C8) acids as the sole source of carbon showed that *C. ingens* can utilize these compounds at all conditions except (1) at 0.1%, pH 4.5 for both compounds and (2) at 0.1%, pH 5.5 for C8. *Pichia mexicana* showed weak growth on 0.05% C6 at a pH of 6.5. Tests conducted with fructose added as a source of carbon indicate the inhibitory effect of the fatty acid salts. In general all yeasts showed some level of growth inhibition. Inhibition increased with decreasing pH and increasing chain length of the compound. The relative order of resistance (most resistant to least resistant) was *C. ingens, P. cactophila, C. sonorensis, P. amethionina* and *P. mexicana.*

TABLE II.   Relative Growth of Representative Organpipe
Yeasts on the Lipids (3%) of Organpipe Cactus
and Two Concentrations (0.05 and 0.1%) of
Caproic and Caprylic Acid. YNB is the Basal
Medium with No Carbon Source, YM is a Complete
Medium and F is 0.5% Fructose (Carbon Source).
0=No Growth, 1=Weak Growth, 2=Moderate Growth,
3=Strong Growth. Tests were Conducted at $25^{\circ}C$
on 2% Agar Plates

Yeast[a]	P.C.	C.S.	P.A.	C.I.	P.M.
YNB	0	0	0	0	0
YNB + Lipids	1	2	1	3	1
YM	3	3	3	3	3
YM + Lipids	3	3	3	3	3
YNB + F					
pH 4.5	3	3	3	1	3
5.5	3	3	3	1	2
6.5	2	2	2	2	2
YNB + 0.05% (C6, C8)[b]					
pH 4.5	0	0	0	2	0
5.5	0	0	0	2	0
6.5	0	0	0	2	(1,0)
YNB + 0.1% (C6, C8)					
pH 4.5	0	0	0	0	0
5.5	0	0	0	(3,0)	0
6.5	0	0	0	(3,2)	0
YNB + F + 0.05% (C6, C8)					
pH 4.5	(1,0)	0	0	(3,2)	0
5.5	(1,0)	(1,0)	0	3	0
6.5	(2,0)	1	(1,0)	3	(1,0)
YNB + F + 0.1% (C6, C8)					
pH 4.5	0	0	0	0	0
5.5	(1,0)	(1,0)	0	(3,0)	0
6.5	(2,0)	(1,0)	(1,0)	3	0

[a] P.C. = Pichia cactophila, C.S. = Candida sonorensis,
P.A. = Pichia amethionina, C.I. = Candida ingens, P.M. =
Pichia mexicana.

[b] Numbers in parentheses depict differences observed for
caproic (C6) and caprylic (C8) acids. C6 is given first.

A. Experiment A

The control vials in this experiment yielded no flies, indicating a microorganism is required for development of *D. mojavensis* on organpipe cactus tissue. These vials were eliminated from further analysis. Table III shows the results for the remaining vials of experiment A. This analysis shows significant effects of the cactus yeasts on time until eclosion, and size of females (males are smaller but show identical response to yeast treatments). Transformed percent emergence showed no effect. The replicates (vials) showed no significant effects for time and size so the observations were pooled for these analyses. Percent emergence is a vial observation and thus constitutes the means for estimating the error. The time effect is due to earlier emergence for yeast treatments *P. mexicana, C. ingens* and *P. cactophila*. The bacterial treatment, and yeast treatments *C. sonorensis* and *P. amethionina* had delayed emergence time. The size of females in this experiment was greatest for *C. ingens* and *P. mexicana* treatments.

B. Experiment B

Table IV shows the adult size results for experiment B. This experiment shows a significant interaction of yeast by lipid treatments. In addition the main effects of yeast, lipids and sex (males are smaller) are all significant. The responses (size) for each yeast treatment can be summarized as follows. *Pichia cactophila* and *P. amethionina* produced similar responses to increased lipid concentration with a depression of size at the highest lipid level. *Candida sonorensis* produced little response with increasing lipids. *Candida ingens* and *Pichia mexicana* produced increased size at 1.5% lipids and a subsequent decrease with increasing lipids in the medium. However, in general *C. ingens* and *P. mexicana* gave greater thorax sizes regardless of the presence or absence of lipids.

TABLE III.   *Analysis of Variance for Experiment A. Obser-*
vations are Time (Days) Until Eclosion,
Female Thorax Size (Ocular Units) and Trans-
formed Frequency of Emergence

Source of Variation	d.f.	S.S.	M.S.	F
TIME				
Corrected total	203	986.63		
Microorganisms	5	143.06	28.61	6.72[a]
Error	198	843.57	4.26	
SIZE (female)				
Corrected total	87	314.18		
Microorganisms	5	101.95	20.39	7.88[a]
Error	82	212.23	2.59	
EMERGENCE				
Corrected total	17	3177.8		
Microorganisms	5	87.8	17.6	0.07
Error	12	2090.0	257.5	

Mean Response and Number per Treatment[b]

	B	P.C.	C.S.	P.A.	C.I.	P.M.
TIME $\bar{X}$	14.67	13.09	14.03	14.24	13.19	12.16
N	36	33	35	37	31	32
SIZE $\bar{X}$	22.73	23.29	22.83	23.18	24.18	25.75
N	15	14	15	17	11	16
EMERGENCE (Per cent) $\bar{X}$	43.7	40.7	42.8	44.6	38.1	40.6
N	3	3	3	3	3	3

[a] $P < 0.001$

[b] B = bacterium, P.C. = Pichia cactophila, C.S. = Candida sonorensis, P.A. = Pichia amethionina, C.I. = Candida ingens, P.M. = Pichia mexicana.

TABLE IV. *Analysis of Variance for Experiment B. Obser-vation is Thorax Length (Ocular Units) for Males and Females*

Source	d.f.	S.S.	M.S.	F
Total	342	1024.52		
Yeast (Y)	5	583.52	72.94	55.92[a]
Lipid (L)	2	17.81	8.90	6.83[a]
Sex (S)	1	315.29	315.29	241.73[a]
Y x L	10	26.70	2.67	2.05[b]
Y x S	5	5.97	1.19	0.92
L x S	2	0.09	0.04	0.03
Y x L x S	9	7.55	0.84	0.64
Error	308	401.73	1.30	

Mean Response for Yeast by Lipid Interaction[c]

		C	P.C.	C.S.	P.A.	C.I.	P.M.
	0%	22.92	25.05	25.46	24.67	25.36	26.25
Lipid	1.5%	21.00	24.48	25.32	25.02	25.90	26.50
	3%	23.75	23.58	25.25	23.64	25.23	25.38

[a] $P < 0.001$

[b] $P < 0.05$

[c] C = control, P.C. = Pichia cactophila, C.S. = Candida sonorensis, P.A. = Pichia amethionina, C.I. = Candida ingens, P.M. = Pichia mexicana.

Table V shows the results for time until eclosion and transformed frequency of emergence on a per vial basis. The main effect (yeast) for time observations is due to earlier emergence from vials with yeasts. The significant yeast effect (emergence observations) is due to lower frequency for control and *C. sonorensis* treatments as compared to frequencies for the other yeasts. The significant lipid treatment effect shows that lipids depress emergence at the highest concentration.

TABLE V.   Analysis of Variance for Experiment B.   Observation is Time (DAYS) to Eclosion and Transformed Frequency of Emergence (EMERGENCE) on
a Per Vial Basis

DAYS

Source	d.f.	S.S.	M.S.	F
Total	35	70.05		
Yeast (Y)	5	65.57	13.11	71.99[a]
Lipid (L)	2	0.33	0.16	0.89
Y x L	10	0.88	0.09	0.48
Error	18	3.28	0.18	

Mean Response for Yeast Treatments[d]

C	P.C.	C.S.	P.A.	C.I.	P.M.
15.83	12.38	11.91	12.17	12.39	12.34

EMERGENCE

Source	d.f.	S.S.	M.S.	F
Total	35	17700		
Yeast (Y)	5	4451	890	3.00[c]
Lipid (L)	2	4031	2015	6.80[b]
Y x L	10	3885	389	1.31
Error	18	5333	296	

Mean Response (Per cent) for Yeast Treatments[d]

C	P.C.	C.S.	P.A.	C.I.	P.M.
24.1	49.2	33.6	55.2	49.9	51.0

Mean Response (Per cent) for Lipid Treatments

0%	1.5%	3%
48.3	53.9	29.23

[a] $P < 0.001$

[b] $P < 0.01$

[c] $P < 0.05$

[d] C = control, P.C. = *Pichia cactophila*, C.S. = *Candida sonorensis*, P.A. = *Pichia amethionina*, C.I. = *Candida ingens*, P.M. = *Pichia mexicana*.

C. Experiment C

The treatment vials containing caprylic acid at levels 2, 3 and 4 yielded no adults. These observations were therefore eliminated from the analysis. The larvae lived for two or more days on level 2 but died within 24 hours at levels 3 and 4. Table VI shows the results for time until eclosion for fatty acid levels 0 and 1 and the yeast treatments. The significant interaction in the analysis of variance is primarily due to the responses observed for *P. amethionina* treatments. In general the no yeast control was delayed in emergence regardless of fatty acid addition. Addition of live yeast speeds up development in the absence of fatty acids. Addition of live yeast speeds up development only in the case of *C. ingens*, when fatty acids are added.

TABLE VI. *Analysis of Variance for Experiment C. Observation is Average Time (Day) Until Eclosion (Per Vial)*

Source	d.f.	S.S.	M.S.	F
Total	17	29.013		
Fatty Acid (F)	1	7.881	7.881	34.23^a
Yeast (Y)	2	13.698	6.849	29.75^a
F x Y	2	4.672	2.336	10.15^b
Error	12	2.763	0.230	

Mean Response for Interaction (F x Y)

	Fatty acid treatment levels	
Yeast treatment	0	1
No yeast	13.55	13.74
P. amethionina	10.69	13.35
C. ingens	11.07	12.20

[a] $p < 0.001$

[b] $p < 0.01$

Table VII shows the results for transformed frequency of emergence for Experiment C. There were no significant effects for this part of the experiment, however, fatty acid addition above the level analyzed in the table was significant (i.e., no flies survived above fatty acid level 1).

TABLE VII. Analysis of Variance for Experiment C. Observation is Transformed Frequency of Emergence (Per Vial)

Source	d.f.	S.S.	M.S.	F
Total	17	1690.24		
Fatty Acid (F)	1	17.72	17.72	0.16
Yeast (Y)	2	350.35	175.18	1.60
F x Y	2	11.01	5.50	0.05
Error	12	1311.16	109.26	

Mean Response (Per cent) for Yeast Treatments

No yeast	P. amethionina	C. ingens
45.10	54.81	54.08

IV   Discussion

The experiments conducted were designed to reveal effects and interactions of yeasts and host plant chemicals on the fitness characteristics of *Drosophila mojavensis*. The results can be related to the natural distributions and frequency of occurrence of the organisms so as to evaluate the likelihood of coadaptation. In general, the major radiations of the family *Drosophilidae* changed habitats according to the similarity of the physiological abilities of the yeasts found in those habitats (Starmer, 1981a). This indicates that co-adaptation is probable and should be based on chemical characteristics of the habitats.

Two major observations led to the experiments presented here. First, organpipe cactus has unusual lipids in its stem tissue (Kircher, 1980, 1982) and second, some yeasts frequently associated with organpipe cactus produce extracellular lipases (Miranda *et al.*, 1982). Table I shows that the lipolytic yeasts *C. ingens* and *P. mexicana*

occur with greatest frequency in organpipe cactus. The
partial specificity of these yeasts supports the notion that
host plant chemistry has an influence on the physiological
abilities of the yeasts. The tests conducted with the lipid
fraction of organpipe (Table II) confirm this in the case of
C. ingens. *Candida ingens* was the best yeast in terms of
(1) using the lipids as a source of carbon, (2) withstanding the free fatty acids and (3) utilizing the free fatty
acids as a source of carbon. This information is also supported by the observations of Henry *et al.* (1976) that *C.
ingens* is naturally enriched on effluent high in C2-C6 fatty
acids. The inhibitory effect of the free fatty acid is known
for other fungi (Wyss *et al.*, 1945). There are two relevant
features of the fungistatic effects of these compounds.
First, the undissociated acid is the active component
(Suomalainen and Oura, 1958) and pH would thus play an important role in the realized activity of these compounds.
Second, activity of saturated fatty acids increases as the
number of carbon atoms in the fatty acid chain increases up
to a point (>C12) where the limited solubility of the acid
restricts its activity. Both the pH effect and chain length
effect are apparent in the cactus yeast experiments (Table II).

Since the fatty acids are mainly combined as esters of
sterol diols and triterpene diols in organpipe cactus, their
inhibitory effect would only be present after release from
their combined forms. Release could be brought about by
extracellular enzymes produced by bacteria and/or possibly
the yeasts. The cactus rotting process starts at a relatively low pH, initially becomes more acidic as a result of
bacterial fermentation, then becomes neutral and finally
basic as the process is completed (Vacek, 1979). The
Drosophila lay their eggs early in the rotting sequence and
thus the larvae should also be exposed to the inhibitory
effects of medium chain fatty acids. It thus seems likely
that yeast capable of withstanding and utilizing the fatty
acids would aid the growth of larvae, either by detoxification of the environment and/or transforming cactus tissue
into larval nutrients.

These notions are supported by the results of experiments A, B and C. Developmental time is shorter for treatment groups of the two lipolytic yeasts and *P. cactophila*.
The observations on size in experiment A are also expected
since development time and size are often negatively correlated (Robertson, 1960).

Unpublished experiments utilizing agria cactus show that
developmental time is equivalent for flies reared from the
yeasts used here (except for *P. amethionina* where development

is longer). Overall the mean developmental time is 11.50
days for flies reared from agria. It is therefore apparent
that developmental time is dependent on both the cactus
tissue and the yeast. In the case of organpipe cactus the
developmental time (Table III) approaches that of flies
reared from agria when the lipolytic yeasts or *P. cactophila*
are present in the rotting tissue.

It should be remembered that Experiment A was conducted
with a bacterium present. The bacterium was utilized since
in general cactus yeasts live principally on products of
primary degradation (i.e. pectinolytic bacteria). A comparison of developmental time of flies reared from bacterial
rotted organpipe or agria with no yeast present emphasize
the importance of bacterial degradation and cactus tissue.
Average time for development on bacterial degraded organpipe
and agria are 14.67 and 11.50 days respectively, which are
significantly different (t = 3.98, d.f. = 50).

The experiment (B) conducted with organpipe lipids added
to banana medium did not include the bacterium. It is thus
possible that all yeasts gave similar developmental rates
since the bacterial component was important in transforming
or degrading the lipids to their active forms (i.e. free
fatty acid). Nevertheless *C. ingens* and *P. mexicana* yielded larger adults when lipids were added (*C. sonorensis* also
yielded large adults). Furthermore all yeasts except *C.
sonorensis* and the control (no yeasts) had relatively high
emergence percentages. Lipids added at concentrations above
the natural level depressed emergence percentage, demonstrating the potency of this chemical fraction.

Experiment C gave results consistent with expectation
based on the results of Table II. That is since *C. ingens*
can tolerate and utilize the free fatty acids, while *P.
amethionina* cannot tolerate the fatty acids, then larvae
should develop faster when *C. ingens* is present. Controls
(no yeast) in this experiment did have some dead yeast added
but the benefit of a live yeast is demonstrated by the control (0 fatty acid) treatment group. As stated in the
results larvae died on fatty acid levels of 2, 3 and 4.
Level 1 is the expected concentration in the rotting cactus.
Death at higher levels confirms the notion that medium chain
fatty acids are a detriment to the larvae.

It is not clear why *P. mexicana* yields benefits to *D.
mojavensis* via an interaction with organpipe lipids. This
is apparent since the yeast shows little ability to utilize
the lipids of organpipe or tolerance to the medium chain
fatty acids Table II. It is possible that more complex

interactions are working. Only more detailed and controlled experiments might reveal the reasons for the *P. mexicana* effect.

V Summary

Results presented here support the idea that the physiological abilities of yeast interact with host plant chemicals to promote the fitness of the insects that serve as dispersal agents for the yeasts. These results are:

(1) Lipolytic yeasts occur most frequently in organpipe cactus.

(2) Organpipe cactus contains relatively large quantities of unusual lipids and medium chain acids.

(3) *C. ingens* utilizes the lipids of organpipe cactus best and the lipids are not in themselves inhibitory to cactus yeasts.

(4) *C. ingens* utilizes and tolerates the medium chain fatty acids better than the other yeasts.

(5) *D. mojavensis* shows increases in fitness characteristics (developmental time and size) when bacteria plus a lipolytic yeast (*C. ingens* and *P. mexicana*) are used as microbial components in the degradation of organpipe cactus tissue.

(6) Organpipe lipids added to banana food above the natural level found in organpipe cactus depress the survival of *D. mojavensis* larvae.

(7) Bacteria are probably important in the beneficial effects of yeasts realized by *D. mojavensis* in organpipe cactus.

(8) Medium chain fatty acids inhibit the growth of *D. mojavensis* larvae.

(9) *C. ingens* serves to detoxify the free fatty acids so that *D. mojavensis* develops faster.

These points provide compelling evidence for the notion that yeasts and *Drosophila* are coadapted with respect to the chemistry of their host plants.

ACKNOWLEDGMENTS

The first stages of this work were started with the collaboration of W.B. Heed and H.W. Kircher. The experiments presented here were ably assisted by Benjamin Metcalf. I

appreciate discussion of the problem with Drs. H.J. Phaff,
D.C. Vacek, W.B. Heed, W.H. Kircher, D.G. Gilbert and J.C.
Fogleman.

REFERENCES

Fogleman, J.C., Starmer, W.T., and Heed, W.B. (1981). *Proc. natn. Acad. Sci. U.S.A. 71*, 4435.
Gilbert, D.G. (1980). *Oecologia 46*, 135.
Heed, W.B. (1978). *In* "Ecology and Genetics: The Interface" (P.F. Brussard, ed.), p. 109. Springer-Verlag, New York.
Henry, D.P., Thompson, R.H., Sizemore, D.J., and O'Leary, J.A. (1976). *Appl. Environ. Microbiol. 31*, 813.
Kircher, H.W. (1980). *Phytochemistry 19*, 2707.
Kircher, H.W. (1982). Chapter 10, this Volume.
Lightle, P.C., Standring, E.T., and Brown, J.G. (1942). *Phytopathology 32*, 303.
Miranda, M., Holzschu, D.L., Phaff, H.J., and Starmer, W.T. (1982). *Int. J. Syst. Bacteriol. 32*, 101.
Phaff, H.J.., and Starmer, W.T. (1980). *In* "Biology and Activities of Yeasts" (F.A. Skinner, S.M. Passmore and R.R. Davenport, eds.), p. 79, Academic Press, New York.
Robertson, F.W. (1960). *Genet. Res. 1*, 288.
Starmer, W.T. (1981a). *Evolution 35*, 38.
Starmer, W.T. (1981b). *In* "Current Developments in Yeast Research" (G.G. Stewart, M. Moo-Young, and I. Russell, eds.), Proc. Fifth Int. Symp. on Yeasts. p. 493. Pergamon of Canada Ltd., Toronto.
Suomalainen, H., and Oura, E. (1958). *Biochim. biophys. Acta 28*, 120.
Vacek, D.C. (1979). Ph.D. Dissertation, Univ. of Arizona.
Wyss, O., Ludwig, B.J., and Joiner, R.R. (1945). *Arch. Biochem. 7*, 415.

12
Interactions Between Microorganisms and Cactophilic *Drosophila* in Australia[1]

Don C. Vacek

Department of Animal Science
University of New England
Armidale, New South Wales

I Introduction

An investigation of the ecological genetics of a holometabolous insect is particularly challenging because the
insect has two radically different life forms. Therefore, a
complete understanding will necessitate studies of the ecological factors that impinge on both. This paper summarizes
the research on one of these factors, the microorganisms of
the breeding sites of cactophilic *Drosophila* in Australia.
Emphasis will be given to the yeasts as components of nutrition influencing potential habitat selection by both adults
and larvae of *Drosophila* species and by *Drosophila* genotypes.
In the last five years, there has been a revived interest
in the analysis of yeast-*Drosophila* relationships. This
interest has been stimulated by the work of Heed *et al.* (1976)
and Starmer *et al.* (1976) on the evolution of cactophilic
Drosophila of the Sonoran Desert of North America. In the
cactus-yeast-*Drosophila* system, relationships can be studied
in detail (Fogleman *et al.*, 1981; Starmer, 1981a, 1981b;
Starmer *et al.*, 1980, 1981) because the breeding and feeding
sites of the fly species are well known (Fellows and Heed,
1972; Heed, 1978).

[1]*Supported by the Australian Research Grants Scheme*

ECOLOGICAL GENETICS AND EVOLUTION
ISBN 0 12 078820 9

In Australia this system has the same advantages as in
the Americas, and the main emphasis of our research is on the
mechanisms maintaining genetic variation at enzyme loci in
natural populations of *D. buzzatii* and *D. aldrichi*, the only
cactophilic species found breeding in *Opuntia* in Australia
(Barker and Mulley, 1976; Barker, 1977; Mulley *et al.*, 1979;
Barker, 1982). Because both species can be found breeding in
the same rot (Mulley and Barker, 1977), the yeasts present in
the rot could be involved in selection occurring both between
and within these two species. Selection for divergence in
yeast feeding and oviposition preference between fly species
is postulated as a way for flies to alleviate interspecific
competition for a common resource, yeast. We also postulate
that genetic polymorphism may be maintained as a result of
selection for divergence in yeast utilization among fly genotypes, to avoid intraspecific competition. We are not asking
the question: Are alleles at a locus neutral or under selection? However, we are asking: Can yeasts act as selection
factors in the maintenance of genetic polymorphisms? There
is strong evidence that at least some of the loci are under
selection (Barker and East, 1980).

An enzyme polymorphism of the fly can be considered an
array of phenotypes, and I am particularly interested in how
the microbial community of the cactus rot might influence
these phenotypic arrays. Cactophilic *Drosophila* are always
in close proximity to rotting cactus material and, in the
larval stage, are in constant direct contact with a microbial
milieu. The myriads of metabolic products of numerous microorganisms may be important to *Drosophila* as, e.g., energy
sources, toxins and sterols. Given that these metabolites
could function as substrates for particular enzymes, and
given that the kinetic properties of various enzyme forms of
a particular locus could vary with different substrates (East,
1982), one can envision the microbial metabolic products
acting as selective factors if the environment is heterogeneous for these metabolites.

In order to elucidate microorganism-*Drosophila* interactions, we are investigating the following three aspects:
(1) microorganism effects at the species level, i.e., adult
feeding and oviposition preferences, larval feeding preferences, and nutritional sufficiency of microorganisms for larval development; (2) microorganism effects at the genotype
level, i.e., association among adult genotypes and their
feeding and oviposition preferences; and (3) heterogeneity of
yeast species in the environment.

II Microorganism Effects at the *Drosophila* Species Level

Yeast preferences of adults and larvae of *D. buzzatii* and (in some studies) *D. aldrichi* have been examined in the field and in the laboratory. Only *D. buzzatii* has been used in experiments investigating the nutritional sufficiency of various microorganisms.

A. *Laboratory Experiments on Adult Feeding and Oviposition Preferences*

Yeast feeding and oviposition preferences of the adult have been examined in the laboratory by direct observation and whole fly assay methods and I shall take some effort to compare these experiments because the implications to future experimental design are quite important.

Barker *et al.* (1981a) used the direct observation method in a series of experiments to test the attractiveness of yeasts from rotting cladodes of *Opuntia stricta* for adults of *D. buzzatii* and *D. aldrichi*. This method involved giving flies a multiple choice of yeast patches in a population cage and counting the numbers of flies and eggs on each yeast at regular intervals throughout one day. A yeast patch was a thick paste of cells harvested from a monoculture in nutrient broth and then spread on an agar disc.

Behavioural differences between sexes, between immature and mature flies and between the two *Drosophila* species were found (Barker *et al.*, 1981a). Regardless of fly and yeast species, significantly more females than males were attracted to yeasts early and late in the day, and oviposition behaviour is the most likely explanation. However, regardless of yeast, the feeding activity of *D. buzzatii* increased as the day progressed, while *D. aldrichi* activity remained relatively constant throughout the day. For immature flies, both species showed significant yeast preferences which were more marked for females. But in no experiment were there significant differences between the two *Drosophila* species in their yeast preferences. One can conclude from the direct observation method that *Pichia cactophila* is commonly the most attractive yeast and *Rhodotorula minuta* var. *minuta* and *Candida sonorensis* are most commonly the least attractive yeasts to both fly species for feeding. No significant oviposition preference was observed within or between species.

I developed the whole fly assay method to eliminate the following shortcomings of the previous method: assumed

correlation of time spent on a yeast patch with number of
cells consumed; confounding of searching, feeding, and oviposition behaviour; and the use of non-actively growing
yeasts harvested from nutrient broth. This improved method
involved giving mature *D. buzzatii* females a multiple choice
of microbial patches for feeding and oviposition over a three
day period, assaying a subset of the flies for yeasts consumed and counting the number of eggs per patch. This method
allows one to actually measure yeasts consumed and to separate feeding from oviposition preference. A patch, which was
renewed at 12 hour intervals, consisted of one of nine microbial communities actively growing on a disc of homogenized
cactus agar medium. This microbial patch, rather than a
paste of microorganisms harvested from nutrient broth, simu-
lates a more natural substrate because cactus-induced metabolites are present. Natural populations of *Drosophila* encounter a microorganism-cactus complex, not microorganisms alone.
A microbial community consisted of a bacterial community
alone or plus one of eight of the common cactophilic yeasts
from *O. stricta* (Table I). The bacterial community was derived from rotting *O. stricta* by elimination of yeasts
through filtration of the rotting material and subsequent
treatment with anti-fungal drugs.

Twenty-seven cages each containing the nine microbial
patches were set up according to a Latin Square design on the
morning of day one, and 30 flies from each of nine cages were
assayed each evening for three days. Table II shows the
ranked mean per cent of colony counts of each yeast per adult.
There is a significant yeast effect on each day. Furthermore,
increasing F ratios for yeast effect from Day 1 to Day 3 indicate that discrimination increases with time. By Day 3,
D. buzzatii females exhibit a clear feeding preference for
P. cactophila and feeding avoidance of bacteria alone, *C.
mucilagina, Rh. minuta* var. *minuta* and *Cryptococcus cereanus*
variety.

The ranked mean number of eggs laid on each yeast (Table
III) shows that *D. buzzatii* oviposits discriminantly, and
discrimination increases with time. By Day 3, the rank order
of preference is effectively identical to that for feeding at
Day 3. These data clearly indicate that *D. buzzatii* oviposits and feeds on identical yeasts.

A comparison of the two methods is most instructive. The
similarities are that *P. cactophila* is usually the most preferred, and *Rh. minuta* var. *minuta* and *C. mucilagina* are
commonly among the least preferred. There are two general
differences between the results of the two experimental
methods: (1) with the direct observation method, there is

TABLE I.   *Microorganism Code and the Per Cent of Total*
Isolates of the 10 Most Common Yeasts in the
Heterogeneity Study (Section IV)

Microorganism	Code	Per cent
Bacteria	Bact.	NA[b]
Candida sonorensis[a]	Cs	27.3
Pichia cactophila[a]	Pc	23.8
Clavispora sp. "O"[a]	ClO	9.5
Pichia sp. "B"[a]	PB	6.5
Rhodotorula minuta var. minuta[a]	Rhm	6.2
Candida mucilagina[a]	Cm	4.3
Pichia opuntiae var. opuntiae[a]	Po	4.3
Cryptococcus albidus var. albidus	Cra	4.3
Pichia amethionina var. amethionina	Pa	4.0
Cryptococcus cereanus variety[a]	Crc	3.0
Other 10 species		6.8

[a] *Yeasts used with bacteria in Section II.A.*
[b] *Not assayed but present (by definition) in every rot.*

TABLE II.   *Feeding Preference as Shown by Ranked Mean*
Per Cent of Colony Counts of Each Yeast[a] in an
Adult for Days 1, 2 and 3 (underlined means
are not significantly different by 5% student-
ized range test)

Day 1 - F = 2.1, P = .05

Rhm	Cm	Po	Bact	Crc	Cs	ClO	Pc	PB
6.2	8.1	8.3	11.6	11.7	12.0	13.4	13.9	14.8

Day 2 - F = 17.8, P < .005

Bact	Cm	Crc	Rhm	Po	PB	Cs	ClO	Pc
0.0	0.3	8.0	8.6	13.8	14.0	15.0	19.8	20.5

Day 3 - F = 31.22, P < .005

Cm	Bact	Rhm	Crc	ClO	PB	Po	Cs	Pc
0.5	1.0	2.4	4.2	10.8	13.6	19.1	20.8	27.6

[a] *See Table I for yeast code*

TABLE III.   *Oviposition Preference as Shown by Ranked*
Mean Number of Eggs Laid on Each Yeast[a] in
the Evening of Days 2 and 3 (underlined means
are not significantly different by 5%
studentized range test)

Day 2 - F = 5.1, P < .005

Crc	Cm	Bact	PB	Po	Cs	Clo	Rhm	Pc
6.0	6.6	12.9	42.6	43.7	48.2	56.9	73.8	80.0

Day 3 - F = 19.3, P < .005

Crc	Cm	Bact	Rhm	Clo	PB	Po	Cs	Pc
21.8	34.0	50.9	87.3	93.7	124.4	187.8	210.1	289.7

[a]*See Table I for yeast code*

less measured discrimination among yeasts as a consequence of significant differences in ranking of yeast species among times within an experiment, and (2) *C. sonorensis* has consistently lower rank than in the whole fly assay method. The first difference is probably due in part to very low metabolic activity of the harvested yeast paste and the short observation period of only one day. The data from the whole fly assay method clearly show that discrimination increases with time. The flies certainly require a period of adjustment to their new cage environment. Furthermore, it is possible that the flies need time to become familiar with the distribution of yeast patches in the cage, just as they might become familiar with the distribution of putative yeast patches in the microcaverns of a rotting cladode or in the distribution of neighbouring rots.

The second difference is probably due in part to confounding of searching behaviour, variation in time for actual consumption of different yeasts (handling time) and actual feeding preference. It is also likely that handling time varies with yeast species because of species differences in colony consistency, cell form, cell aggregation characteristics and extracellular polysaccharide production. This handling time factor coupled with search behaviour could destroy any correlation between time spent on a yeast patch and amount of yeast consumed. In fact, Barker *et al.* (1981a) assayed the yeasts in crops of flies from one cage of one experiment and found marked differences between the distribution of flies attracted to yeast patches and the distribution of yeasts in these flies. *Candida sonorensis*, in particular, was more frequent in the crops than expected from feeding preference ranking. Furthermore, potential differences in digestibility between yeasts can influence the outcome of the whole fly assay method. However, Fogleman *et al.* (1981) compared the rate of digestion of two differentially preferred yeasts, *P. cactophila* and *Clavispora* sp. "0", in larvae of *D. mojavensis* and found no difference.

B. *Field Experiments on Yeast Attractiveness to Adults*

Three experiments designed to determine feeding preference of the two fly species in the field were performed by Barker *et al.* (1981b) over an 18 month period. The number of flies on each yeast bait was recorded at regular time intervals; a bait consisted of a single yeast species growing on homogenized cactus. *Drosophila buzzatii* showed highly significant differential attraction to the yeast baits, but the relative attractiveness of the yeasts was radically different among the experiments which were performed in different seasons. They found weak indications that these changes in attractiveness among experiments could be related to seasonal differences in the relative frequencies of the yeasts in natural rots. No evidence was found for differences in the attractiveness of yeasts for sexes in *D. buzzatii*, as was found in the laboratory experiments (Barker *et al.*, 1981a). Only in one experiment was *D. aldrichi* in sufficient frequency to be compared with *D. buzzatii* for yeast preference. They found highly significant differences between the species in their yeast preferences, thus indicating a degree of niche separation.

C. *Larval Preference in the Laboratory*

Larvae of *D. buzzatii* and *D. aldrichi* were given a choice of nine yeasts presented in 36 pairwise combinations per replicate. Each pair was presented as four 1.5 cm diameter yeast patches actively growing on homogenized cactus agar in a petri dish; patches of like species were diagonal to each other. The number of larvae on each yeast patch was recorded at 5, 15, 30, 45 and 60 minutes and these data were used to construct for each yeast per replication an attractivity index (Lindsay, 1958). Preliminary analyses of the 45 and 60 minute recordings provide nearly identical ranking of attractivity indices. Table IV shows the ranked mean attractivity indices over four replications at 60 minutes. Clearly,

TABLE IV.   *Larval Feeding Preference as Shown by Ranked*
Mean Attractivity Indices for D. buzzatii and
D. aldrichi for Each Yeast[a] at 60 Minutes
(underlined means are not significantly dif-
ferent by 5% studentized range test)

D. buzzatii - F = 22.1, P < .001

Crc	Rhm	Po	Cm	PB	Pc	Cs	Cra	ClO
0.22	0.31	0.43	0.45	0.58	0.58	0.63	0.63	0.66

D. aldrichi - F = 28.1, P < .001

Crc	Rhm	Cm	Po	PB	Pc	Cra	Cs	ClO
0.14	0.30	0.41	0.48	0.51	0.62	0.64	0.69	0.72

[a]*See Table I for yeast code*

larvae of each species do selectively feed; however, there is
no significant difference between *Drosophila* species in yeast
preference.

The yeast preferences of *D. buzzatii* larvae (Table IV)
and adults (Table II) are very similar. With the exception
of *P. opuntiae* var. *opuntiae* and the treatments not common to
both experiments, viz., bacteria and *Cr. albidus* var. *albidus*,
the preferred group represented by the top four yeasts and
the non-preferred group represented by the remaining three
yeasts are identical for both larvae and adults. Differences
are that *P. opuntiae* var. *opuntiae* is more preferred by
adults than by larvae, and adults have a clearly preferred
yeast, *P. cactophila*. Larvae have no single preferred yeast
but a group of five preferred yeasts.

D. Nutritional Sufficiency of Microorganisms

Thirty axenic larvae were placed into food vials with 7
ml of cactus homogenate supporting actively growing yeasts
and bacteria (treatments). There were 18 treatments (replicated 12 times) as follows: eight yeasts each alone, eight
yeasts plus the bacterial community (as described in Section
II.*A*), the bacterial community alone, and no microorganisms
(cactus control). Developmental time (A), i.e., mean number
of days from egg to emergence; per cent of larvae developing
to emergence (B); and a developmental index (the ratio B/A)
were calculated as parameters of larval growth. It is clear

that the treatments are not nutritionally equivalent because the microoganism effect is highly significant (P < .001) for all three parameters (Table V). Bacteria alone or bacteria plus any one of the eight yeasts gives shortest developmental time, and *C. mucilagina, C. sonorensis,* and *Clavispora* sp."0" give second shortest developmental time. All larvae died in cactus control. Per cent emergence shows more overlap and slightly different ranking of treatments, e.g., *C. mucilagina* shows short developmental time but a low per cent emergence.

However, the developmental index is a more meaningful parameter of larval growth because it incorporates both of the emergence parameters, per cent emergence and developmental time, which are important fitness components (Lewontin, 1965; Powell and Taylor, 1979). In organisms such as cactophilic *Drosophila* which encounter fluctuating resources (see Sections IV and V), the index will be positively correlated

TABLE V. *The Effect of Bacteria and Yeasts[a] on the Nutritional Sufficiency of the Host Cactus as Shown by Mean Larval Growth Parameters Ranked by Increasing Nutritional Value (underlined means are not significantly different by 5% studentized range test)*

A. *Mean number of days from egg to emergence (A):*

Rhm	Po	PB	Crc	Pc	ClO	Cs	Cm	Bact. + / - yeast
28	24	21	20	18	17	16	16	15

B. *Mean per cent developing to emergence (B):*

Po	Rhm	Cm	PB	Cs	ClO	Crc	Bact. + / - yeast	Pc
63	66	72	81	83	84	84	86	88

C. *Mean developmental index (B/A):*

Rhm	Po	PB	Crc	Cm	Pc	ClO	Cs	Bact. + / - yeast
2.4	2.7	3.8	4.2	4.7	4.8	5.0	5.1	5.9

[a]*See Table I for yeast codes*

with fitness. The index is highest for bacteria alone, or
plus any one of the yeasts, and second highest for a group of
three yeasts, *C. sonorensis*, *Clavispora* sp. "0", and *P.
cactophila*. *P. opuntiae* var. *opuntiae* and *Rh. minuta* var.
minuta are clearly the yeasts on which the larvae would have
lowest fitness.

Some interesting qualitative observations were made on
the cactus control. At 20 days there were active first instar larvae present but only a very small percentage of the
original 30 larvae. If at this stage bacteria, yeast, and
water were added, a small proportion of these remaining larvae developed into adults. Thus *D. buzzatii* can suspend
development for some period of time until nutritional and
osmotic conditions improve. This adaptation could be quite
important under stress conditions when microbial growth is
drastically reduced due to temporary dessication of the rot.
Furthermore, the adults are relatively dessication resistant
(Parsons and McDonald, 1978; Gunawan and Barker, 1980), an
important characteristic of a colonizing species, and *D.
buzzatii* certainly is a successful colonist from South
America.

Microbial characteristics that affect nutritional value
of a substrate for *Drosophila* can be divided into two groups,
viz., growth rate and substrate transformation. The latter
refers to the metabolic properties of the microorganism that
affect cell composition, cell digestibility, depletion of
substrate nutrients or toxins, and release or production of
nutrients and toxins from substrate macromolecules. The
above statement is a very strong argument for the use of host
plant as the basal medium in experiments assessing attractiveness (see Section II.*A*) or nutritional value of microorganisms for *Drosophila*, because in the natural environment,
adults and larvae respond to a host plant-microbe complex.
Kearney and Shorrocks (1981) have clear evidence that the
chemical composition of the basal growth medium affects the
nutritional adequacy of yeasts for three species of European
temperate *Drosophila*.

It is clear that bacteria growing on *Opuntia* produce the
fastest larval development which is not altered by the addition of yeasts. Differences in growth rate among the yeasts
and bacteria could be one factor contributing to slower larval development on yeasts alone. *Pichia opuntiae* var.
opuntiae, *Rh. minuta* var. *minuta*, and *Cr. cereanus* variety,
which are low on the developmental index scale, grow dis-
tinctively slower on homogenized cactus than the other
yeasts; whereas, bacteria, which are highest on the scale,
have an intrinsically faster growth rate.

There are striking similarities between the developmental index (Table V) and larval preference of *D. buzzatii* (Table IV). *Pichia cactophila, Clavispora* sp. "0" and *C. sonorensis*, the most nutritious yeasts, are in the group most preferred by larvae, whereas *Rh. minuta* var. *minuta* and *P. opuntiae* var. *opuntiae*, the least nutritious yeasts, are very low on the larval preference scale. Thus larvae have the ability to forage and select the yeasts on which they have highest fitness.

III   Microorganism Effects at the *Drosophila* Genotype Level

Yeast feeding preferences of genotypes of *D. buzzatii* have been examined in the field (Barker *et al.*, 1981b) and in the laboratory. Oviposition preferences of genotypes, as measured by the genotypes of the eggs laid, have been studied in the laboratory.

Drosophila buzzatii attracted to yeast baits in the field as described in Section II.*B* were assayed electrophoretically on starch gels for six loci known to be polymorphic. Chi-square heterogeneity tests for association of genotypes and alleles with yeasts showed some evidence for association at the *Esterase-2* locus. Barker *et al.* (1981b) concluded that this differential attraction of genotypes to yeasts suggests that nutritional variation in the form of yeasts may contribute to within population selection and the maintenance of allozyme polymorphisms.

In the laboratory, adults assayed for yeast in the experiment involving the whole fly assay method (Section II.*A*) also were analysed electrophoretically for seven loci known to be polymorphic in the field population from which these adults were derived. For each locus a Chi-square heterogeneity analysis was done on the number of each genotype and each allele associated with each yeast consumed by the fly. There were no significant values so we must conclude that the flies feed randomly with respect to their genotype. *Esterase-2* was the only locus close to significance (P = .10).

A subset of the eggs scored on each yeast on days 2 and 3 under the whole fly assay method (Section II.*A*) were reared to adults under low density conditions on standard laboratory medium. The emergences were assayed electrophoretically for the seven loci. Chi-square heterogeneity analyses showed highly significant (P < .001) yeast effect for genotypes at *Esterase-1, Esterase-2*, β-*N-acetylhexosaminidase* and *Leucine-aminopeptidase-1*; significant (.025 < P < .05) yeast effect for *Phosphoglucomutase* and *Alcohol dehydrogenase-1*; and no

yeast effect for *Aldehyde oxidase*. Results for alleles were
identical with one exception, *Aldehyde oxidase*, which was
significant (P = .05). These data strongly suggest that
D. buzzatii females having different genotypes oviposit on
yeasts in a non-random manner. However, the data are not
conclusive because the correlation between fly genotype and
egg genotype is not unity, even if one assumes complete
assortative mating.

IV   Heterogeneity of Yeasts in the Environment

How does rot age, rot type, microclimate or site, and
season affect yeast species distribution? Over a 15 month
period at one locality in the Hunter Valley, N.S.W. (locality
5 of Barker and Mulley, 1976), 370 yeast isolates representing 20 species were collected from 278 rots. The concentration of each yeast per sample was not ascertained; therefore,
frequency of isolation is used as a measure of abundance.
The average number of species per rot was 1.27 ± 0.082
($\overline{X}$ ± SE) and 20% of the rots had no yeast. *Candida sonorensis* and *P. cactophila*, the most common yeasts, represented
51% of the isolates. The next eight most common yeasts made
up 42% of the isolates (Table I). The rots were defined in
terms of the affected plant part, viz., young cladode, old
cladode, basal cladode or basal stem. Young and old cladodes
are both aerial on the plant, but the former is current season growth and the latter is of the previous year or older.
Basal cladodes lie on the ground and are usually rooted, old,
and often partly covered by leaf litter and other cladodes.
Basal stems are the main "trunk" of the plant and the rots
may extend from the above ground section to 10 cm or more
below ground. These rots were collected at ten sites (approximately 60-70 m apart), five sites in each of two parallel
and linear south-west to north-east transects. One transect
ran along the crest of a ridge and the other along the
southern slope. Based on exposure and temperature these
sites represent different microhabitats.

There were significant differences among seasons and rot
types in average number of yeast species per rot. Yeast
numbers were low in summer when rot duration is measured in
weeks and were high in basal stem rots which may last for
some months. These differences in average number of yeasts
per rot are supported by analysis of the distribution of the
10 most common yeast species, which show temporal effects to
be the major component contributing to heterogeneity;

microgeographic effects based on site and rot types were less important but, nevertheless, significant.

Several important observations can be made on individual yeasts. As expected, the most common yeasts, *C. sonorensis* and *P. cactophila,* were present in all months and had similar frequencies in each rot type. The other yeasts were more restricted over time. *Pichia opuntiae* var. *opuntiae,* although only 4% in overall frequency, reached its highest frequency in spring when approximately 63% of the *P. opuntiae* var. *opuntiae* isolates were obtained. *Clavispora* sp. "0", although only 9% in overall frequency, reached its highest frequency in young cladodes on the southern slope in spring when approximately 70% of the *Clavispora* sp. "0" isolates were obtained and when it is similar in frequency to *C. sonorensis* and *P. cactophila.*

The frequency of rots without yeasts was significantly different among seasons and rot types. There were high frequencies of no yeast rots in the summer and in old and young cladode rots. At the time of sampling adults were active on 19% of the rots without yeasts and adults were reared from some of these rots.

V   *Drosophila* Preferences in the Light of Yeast Heterogeneity in the Host

Are the preferences exhibited by the adults and larvae ecologically relevant, i.e., can the yeast preference eventuate in a natural rot? The data show the environment to be significantly heterogeneous in yeast distribution among seasons and rot types. Therefore, the possibility exists for *Drosophila* to selectively feed and oviposit on yeasts in a natural rot. However, a larva or adult foraging in a rot will rarely have a choice of three or more yeasts because an average rot contains only one to two yeast species.

Most drosophilologists assume that yeasts are absolutely necessary for both adult and larval nutrition. However, the nutritional sufficiency data show that yeasts are not necessary and that bacteria are best for larval growth. But it is highly likely that bacteria occur in several combinations in a rot, and we used only one of these combinations of unknown bacteria. If other bacteria combinations are not as nutritious, yeast must play a role in nutrition of *Drosophila*. Furthermore, if bacteria are most important, larvae should prefer them over yeasts; this choice experiment remains to be done. Nevertheless, the sufficiency of bacteria alone explains why adults have been reared from rots without detectable yeasts.

One must not ignore the most elementary explanation for
adult preferences - yeasts preferred by adults are simply
indicators of a suitable breeding site, i.e., one which contains bacteria. A preliminary chromatographic analysis (East,
unpublished data) of volatile compounds produced by four
yeasts growing in pure culture on homogenized cactus showed
that *P. cactophila* and *C. sonorensis*, which are the most
common yeasts and the ones most preferred by adults, produce
approximately twice as much of the alcohols, and the esters
derived from them, as do *P. opuntiae* var. *opuntiae* and *Cr.
cereanus* variety. The two former yeasts could be used as a
cue for an active microbial community which always includes
bacteria. In natural populations during stress conditions a
fly must be efficient at locating a relevant microbial community. It may be no coincidence that the most common yeasts
are the most aromatic, the most attractive to adults and
relatively nutritious for larvae; these relationships may be
a result of coevolution between *Drosophila* as vectors for
yeasts and yeasts as indicators of good breeding sites for
Drosophila.

Nevertheless, this hypothesis that preferred yeasts are
simply indicators of a suitable breeding site does not pre-
clude optimal foraging by adults and larvae. The cactus rot
as a resource is quite microbially heterogeneous. Adults can
feed and oviposit on an optimal rot which may remain optimal
for several weeks or which a week later may become suboptimal
as the microbial population changes in composition and size;
O. stricta rots can be persistent in the cooler seasons or
small and ephemeral in the summer. Therefore, the larvae
also should be expected to optimally forage in the rot in
order to complete development before the rot becomes unsuit-
able.

Such a heterogeneous microbial environment provides a
potential for habitat selection and diversifying selection to
contribute to the maintenance of genetic variation in *D.
buzzatii* (Barker, 1982). The highly significant associations
of genotypes and alleles with yeasts in some of the field and
laboratory experiments (Section III) are very encouraging.

VI   Future Research

The heterogeneity study gives us an idea of the spatial
distribution of yeast that an adult might encounter among
rots. However, it tells us nothing about the patchiness of
yeasts in time within a rot. An investigation of the within
rot distribution of yeasts is particularly important for an

understanding of the foraging behaviour of larvae which are restricted to a single rot. A comparison of the frequency of yeasts in a section of a rot with the frequency of yeasts in the larvae feeding in that section will allow us to test the hypothesis that *Drosophila* do discriminate among yeasts in a natural rot. Monitoring of yeasts in time gives information on yeast population flux, which affects the change in choice of yeasts a larva may have during development. Concurrent assaying of genotypes will allow us to detect genotype-yeast associations in natural populations of *D. buzzatii*.

Laboratory investigations do not support the hypothesis that *D. buzzatii* and *D. aldrichi* co-exist by specializing on different yeasts. However, Barker *et al.* (1981b) have evidence that these species have different yeast preferences in the field. Future research will focus on a comparison of gut analysis of adults of both species in natural populations and a comparison of the yeast flora of sections of rots with the species that eclose from those sections.

Bacteria can be sufficient for larval development; therefore, the bacterial component of the system can no longer be neglected. We have only begun to appreciate their diversity (Young *et al.*, 1981) and the future challenge is to recognize key species and understand their importance in the cactus-yeast-*Drosophila* system.

ACKNOWLEDGMENTS

I am very grateful to Professor J.S.F. Barker, Peter East, Hani Soliman and Annette Kennewell for their contributions to these studies. Data used in the discussion on heterogeneity of yeasts in the environment was taken from a manuscript by J.S.F. Barker, G.L. Toll, P.D. East, M. Miranda and H.J. Phaff.

REFERENCES

Barker, J.S.F. (1977). *In* "Measuring Selection in Natural Populations" (F.B. Christiansen and T. Fenchel, eds.), Lecture notes in Biomathematics *19, 403*. Springer-Verlag, Berlin.
Barker, J.S.F. (1982). Chapter 14, this Volume.
Barker, J.S.F., and East, P.D. (1980). *Nature, Lond. 284*, 166.
Barker, J.S.F., and Mulley, J.C. (1976). *Evolution 30*, 213.

190   *Don C. Vacek*

Barker, J.S.F., Parker, G.J., Toll, G.L., and Widders, P.R.
    (1981a). *Aust. J. Biol. Sci. 34*, 593.
Barker, J.S.F., Toll, G.L., East, P.D., and Widders, P.R.
    (1981b). *Aust. J. Biol. Sci. 34*, 613.
East, P.D. (1982). Chapter 21, this Volume.
Fellows, D.P., and Heed, W.B. (1972). *Ecology 53*, 850.
Fogleman, J.C., Starmer, W.T., and Heed, W.B. (1981). *Proc.
    natn. Acad. Sci. U.S.A. 78*, 4435.
Gunawan, B., and Barker, J.S.F. (1980). *Drosophila Inf. Serv.
    55*, 53.
Heed, W.B. (1978). *In* "Ecological Genetics: The Interface"
    (P.F. Brussard, ed.), p. 109. Springer-Verlag, New York.
Heed, W.B., Starmer, W.T., Miranda, M., Miller, M.W., and
    Phaff, H.J. (1976). *Ecology 57*, 151.
Kearney, J.N., and Shorrocks, B. (1981). *Biol. J. Linn. Soc.
    15*, 39.
Lewontin, R.C. (1965). *In* "The Genetics of Colonizing Species"
    (H. Baker and G.L. Stebbins, eds.), p. 77. Academic Press,
    New York.
Lindsay, S.L. (1958). *Am. Nat. 92*, 279.
Mulley, J.C., and Barker, J.S.F. (1977). *Drosophila Inf. Serv.
    52*, 151.
Mulley, J.C., James, J.W., and Barker, J.S.F. (1979). *Biochem.
    Genet. 17*, 105.
Parsons, P.A., and McDonald, J. (1978). *Experientia 34*, 1445.
Powell, J.R., and Taylor, C.E. (1979). *Am. Scient. 67*, 590.
Starmer, W.T. (1981a). *Evolution 35*, 38.
Starmer, W.T. (1981b). *In* "Current Developments in Yeast
    Research" (G.G. Stewart, M. Moo-Young, and I. Russell,
    eds.), Proc. Fifth Int. Symp. on Yeasts, p. 493.
    Pergamon of Canada Ltd., Toronto.
Starmer, W.T., Heed, W.B., Miranda, M., Miller, M.W., and
    Phaff, H.J. (1976). *Microbial Ecol. 3*, 11.
Starmer, W.T., Kircher, H.W., and Phaff, H.J. (1980).
    Evolution 34, 137.
Starmer, W.T., Phaff, H.J., Heed, W.B., Miranda, M., and
    Miller, M.W. (1981). *Evol. Biol. 14*, 269.
Young, D.J., Vacek, D.C., and Heed, W.B. (1981). *Drosophila
    Inf. Serv. 56*, 165.

13
The Role of Volatiles in the Ecology
of Cactophilic *Drosophila*

James C. Fogleman

Department of Ecology & Evolutionary Biology
University of Arizona
Tucson, Arizona

Most of the species in the genus *Drosophila* and all of
the cactophilic species are saprophytes. Thus, the bacteria,
yeasts, and molds which grow on decaying plant material
represent major components of the food resources for both
larval and adult stages of the life cycle of these insects.
The human appreciation of the main metabolic product of certain yeasts precedes recorded history. More recently, other
products of microbial metabolism, such as lower molecular
weight alcohols and their acetates (often referred to as
fusel oils), have been investigated. A few of these studies
involved the effect of these volatile compounds on insect
behavior. It is the primary purpose of this paper to examine
the cactus-microorganism-*Drosophila* model system in reference to the possible role(s) of volatile compounds in the
ecology of cactophilic *Drosophila*.

I Yeasts and Their Volatiles

The fermentation of sugars to ethanol by yeasts is a
process which has drawn attention for centuries. With the
development of more sophisticated analytical techniques,
other metabolic products of yeasts (and other fungi) have
also been identified. Some of the more common volatiles
identified as yeast metabolites are: ethanol, ethyl acetate,
1-propanol, *iso*-butanol, *iso*-amyl alcohol, and optically
active amyl alcohol (Phaff *et al.*, 1978). Other volatiles

ECOLOGICAL GENETICS AND EVOLUTION
ISBN 0 12 078820 9

produced by fungi (including molds) which have been reported, but usually occur in lower concentrations, include: 2-propanol, 1-butanol, 2-butanol, 1-hexanol (Ingraham and Guymon, 1960); methyl acetate (Tabachnik and DeVay, 1980); 2,3—butanediol (Phaff *et al.*, 1978); acetic acid, acetone, acetaldehyde, 2-butanone, diacetyl (Prior *et al.*, 1980; Gyosheva and Rusev, 1979); furfuraldehyde, 2-hexanal, 2-methyl butanal (Collins and Kalnins, 1965); and isobutyl acetate, isoamyl acetate, methanol (Collins and Morgan, 1960).

The filamentous fungi (molds) which live in cactus necroses have not been extensively studied. The yeasts, on the other hand, have been well characterized both physiologically and taxonomically (see Starmer *et al.*, 1982 for an excellent review). Previous reports of volatiles produced in cactus rots stated that only three species of cactophilic yeasts can ferment glucose: *Torulopsis sonorensis* (now called *Candida sonorensis*), *Candida tenuis* (now called *Pichia mexicana*), and *Pichia cactophila* (Vacek, 1979; Young *et al.*, 1981). The only volatile reportedly produced in fermentation tests was ethanol. These tests, however, were performed under fairly anaerobic conditions. Phaff *et al.* (1978) point out that the by-products of yeast metabolism can vary greatly both qualitatively and quantitatively under anaerobic and aerobic growth conditions, and surprisingly high concentrations of by-products (other than ethanol) have been found when yeasts are grown under aerobic conditions. With this in mind, the volatiles produced by the cactophilic yeasts were re-examined.

Flasks containing a complete liquid medium (YM broth) were inoculated with a single yeast species from a 24 hour culture and agitated vigorously on a mechanical shaker at 25°C. Samples (0.5 ml.) were taken at 24, 48, and 96 hours after inoculation, centrifuged, and stored in a freezer until analyzed by gas chromatography.

The resulting volatile profiles are given in Table 1. A total of five different compounds were detected in concentrations high enough to measure ($\geq$ 0.1mM): methanol, ethanol, ethyl acetate, *iso*-butanol, and *iso*-amyl alcohol. For a single strain, a maximum concentration of volatiles in the medium usually occurred in the 48 hour sample. Differences between species in volatile concentrations were independent of cell density. It is evident from the data in Table I that qualitative and quantitative differences between the yeast species exists for all five volatiles. Methanol was not produced in high concentration by any yeast, but half of the cultures had no methanol at all. In addition to the

TABLE I. Maximum Concentrations of Volatiles Produced by Cactophilic Yeasts in YM Broth

Species	Strain number	Log (cells/ml)	Molarity (mM)				
			MeOH	EtOH	EtOAc	iso-BuOH	iso-Amyl
Pichia cactophila	80-143.1	8.342	---	4.5	2.4	0.3	0.3
	80-243A	8.322	---	13.2	1.8	0.3	0.2
Pichia amethionina v.a.	81-228.4	8.176	0.1	34.8	10.9	0.4	0.2
	81-265.1	8.415	0.3	50.1	8.5	0.4	0.2
Pichia amethionina v.p.	79-227.1	8.415	---	3.5	4.8	0.3	---
	80-314.1	7.946	---	1.7	5.9	1.4	0.7
Pichia amethionina	81-62	8.415	0.9	41.9	7.3	0.3	0.1
Pichia mexicana	80-124.1	8.114	---	36.2	---	0.4	0.2
Pichia opuntiae	76-211	8.602	---	---	---	---	0.2
	76-385B	8.792	---	---	---	0.2	---
Pichia heedii	78-20	8.204	---	0.3	---	---	---
	80-312.8	8.114	0.2	0.2	---	0.2	0.2
Pichia species "A"	76-296B	6.973	0.3	16.2	---	0.2	0.3
	78-98	7.846	0.1	19.5	---	0.3	0.4
Candida sonorensis	71.148	7.651	0.2	67.7	---	0.4	0.4
	78-34	7.974	0.7	52.0	---	0.7	0.6
Candida ingens	80-312.9	8.176	---	2.2	---	---	---
	80-143.2	7.403	0.1	0.1	---	---	---
Candida mucilagina	78-102	7.845	---	0.3	---	---	---
Candida species "K"	76.242A	7.185	---	0.1	---	---	0.1
	78-101	8.037	---	---	---	---	0.1
Kluyveromyces marxianus	80-79-5	8.398	0.1	59.4	2.4	0.4	0.6
Cryptococcus cereanus	79-259.1	8.813	0.2	1.2	---	0.2	0.1
	80-322.6	8.519	0.1	0.9	---	0.2	0.2
Clavispora species "O"	79-245.3	8.602	---	13.5	---	0.2	---
	79-290.1	8.708	0.2	32.1	---	0.2	0.1

three species previously mentioned, ethanol was produced in significant concentrations by *P. amethionina* variety *amethionina*, *P.* species "A", *K. marxianus*, and *Cl.* species "O" and to a lesser extent by *P. amethionina* variety *pachycereana*, *C. ingens*, and *Cr. cereanus*. In fact, only three out of the 26 strains tested did not show a detectable level of ethanol. Ethyl acetate, synthesized by the enzymatic esterification of ethanol and acetic acid, was made by *P. cactophila*, *P. amethionina* (both varieties), and *K. marxianus*. *Iso*-butanol and *iso*-amyl alcohol were detectable in most of the cultures, but only three yeasts, *C. sonorensis*, *P. amethionina* v. *p.*, and *K. marxianus*, stand out as producing relatively more than the others. Several other volatiles (i.e. 1-propanol, 2-propanol, *iso*-butyl acetate, and optically active amyl alcohol) were sometimes detectable but less than 0.1mM in concentration.

The disparity between the above results and those previously reported for cactophilic yeasts is most likely due to the differences in by-product formation under aerobic versus anaerobic conditions. Whether these yeasts contribute significantly to the volatile profiles observed in samples of naturally occurring cactus rots remains to be determined. It is important, however, to point out that their putative contribution would probably be rather dynamic. That is, yeasts can consume some volatile compounds as well as produce them, and, in the case of certain volatiles, will do both. In one time course study, a culture of *P. amethionina* v. *p.* was sampled every four hours and ethyl acetate was found to increase in concentration up to between 32 and 36 hours at which time it disappeared completely from the culture medium (Fogleman, unpubl.). All of the cactophilic yeasts can use ethanol as a carbon source. Methanol, 2-propanol, acetone, and ethyl acetate can be used by some species and not others. Whether a particular yeast is adding volatiles to the substrate or subtracting them presumably depends on the availability of usable sugars. Finally, yeast communities in natural rots are thought to undergo a species succession which may also affect the types and amounts of volatiles present (Starmer, 1982).

II Bacteria and Volatiles

Little work has been done on the resident bacteria of cactus rots. It is obvious that bacterial action must precede yeast growth since all but one of the yeasts

(*K. marxianus*) listed in Table I lack the enzymes necessary
to break down plant cell walls (cellulase, hemicellulase, or
pectinase). Bacteria, on the other hand, do possess this
ability. *Erwinia carnegieana* is one bacterial species which
is almost always found in association with soft rot in
saguaro (Lightle *et al.*, 1941; Boyle, 1949; and Alcorn and
May, 1962), and can be isolated from rots in other species of
columnar cacti as well. This bacterium does not appear to be
responsible for causing cactus disease since rot formation
does not typically occur when healthy plants are injected (or
injured slightly and inoculated) with *Er. carnegieana*
(Steenbergh, 1970). It is possible that the pH of healthy
cactus tissue (approx. 5.0) is too low to permit bacterial
growth (Boyle, 1949). In addition to *Erwinia*, seven other
genera were isolated from saguaro and cardón (*Pachycereus
pringlei*) rots by Graf (1965), and eight genera were found
in samples of agria (*Stenocereus gummosus*) and organ pipe
(*S. thurberi*) rots by Young *et al.* (1981). The majority of
these bacteria belong to the family Enterobacteriaceae.

Interest in the volatiles produced by bacterial action on
cactus tissue was prompted by the observation that some
cactus necroses with little or no yeasts present attracted
large numbers of adult *Drosophila* and were used as oviposition sites (Starmer, 1982). Previous studies of the
volatiles produced by enterobacteria in general and by
Erwinia species in particular report that the usual end-
products of glucose fermentation are a combination of some
or all of the following: acetic, formic, succinic, and
lactic acids, ethanol, CO_2, acetoin (2-hydroxy-3-butanone),
and 2,3-butanediol (White and Starr, 1971; VanVuure *et al.*,
1978). Other volatiles reportedly produced by enterobacteria
include: methanol, acetone, 2-butanone, 3-pentanone,
heptanal, and several unidentified carbonyl compounds
(Thayer and Ogg, 1967).

In order to identify the volatiles which result when
Er. carnegieana is grown on cactus tissue, the following
experiment was performed. Glass Petri plates containing
approximately 25 grams of homogenized-autoclaved cactus
tissue were inoculated with one of two strains of *Er.
carnegieana*. These two strains had been originally isolated
from agria and organ pipe and were obtained from S. Alcorn
(University of Arizona). Duplicate plates of five cactus
species, agria, organ pipe, saguaro, senita, and prickly
pear (*Opuntia*), were inoculated with 2 ml. of a suspension
of bacteria. Samples were taken at 24, 48, and 72 hours
after inoculation, centrifuged, and frozen until analyzed
by gas chromatography.

The volatile profiles for *Erwinia* rotted cacti are shown in Table II.   The maximum concentration of methanol ranged from 2 to 5 mM in all plates.   Methanol, however, is not a fermentation endproduct of the bacteria, but rather is liberated as a result of the enzymatic digestion of the pectin in plant cell walls (Schink and Zeikus, 1980).   For this reason, it was not surprising to find that the cacti were fairly equal with respect to methanol formation.   It was very surprising, however, to find that methyl acetate was

TABLE II.   *Maximum Volatile Concentration Achieved Within*
3 Days when Cacti are Rotted with Erwinia
carnegieana.   No Volatiles were Detected in
Unrotted (Control) Samples

Cactus	Strain origin[a]	Molarity (mM)				Rel. pk. area
		MeOH	MeoAC	EtOH	EtoAc	Acetoin
Agria	A	2.3	0.0	1.1	1.9	1.3
	A	2.7	0.0	2.0	1.4	1.0
	OP	2.5	0.0	0.5	0.0	0.5
	OP	1.7	0.0	0.3	0.1	0.6
Organ pipe	A	2.6	0.0	4.8	6.7	0.9
	A	2.7	0.0	4.8	4.1	1.2
	OP	1.7	0.2	4.2	3.9	0.6
	OP	1.9	0.0	2.7	2.0	0.1
Saguaro	A	5.3	0.9	0.0	0.7	9.6
	A	3.3	0.7	0.0	0.3	5.2
	OP	4.3	0.7	0.0	0.3	4.2
	OP	3.3	0.9	0.1	0.0	4.9
Senita	A	3.7	0.1	2.9	4.2	1.8
	A	3.9	0.0	3.0	1.8	1.1
	OP	4.3	0.0	2.9	0.0	1.7
	OP	4.0	0.0	2.4	1.1	0.8
Opuntia	A	4.5	0.0	2.4	1.0	6.5
	A	4.3	0.0	3.1	2.5	5.1
	OP	3.6	0.0	3.9	2.8	8.3
	OP	3.3	0.0	2.6	1.9	4.8

ANOVA-Prob of Sig. Diff. Between:

Cacti	---	<0.001	<0.01	n.s.[b]	<0.01
Er. strains	---	n.s.	=0.1	<0.1	n.s.

[a] *A=agria, OP=organ pipe (UA strain numbers 78-28 & 66-78).*

[b] *n.s. = not significant*

produced in significant amounts only when saguaro was the substrate. Methyl acetate, then, represents a qualitative difference between the cacti. Saguaro again stands out when maximum concentrations of ethanol are compared because virtually none was produced in the saguaro plates. The maximum concentration averaged over strains and replications showed that more ethanol was present in the organ pipe plates ($\bar{X}$ = 4.1 mM) than in *Opuntia* (3.0), senita (2.8), or agria (1.0). This pattern of concentrations between cacti is repeated with respect to ethyl acetate. Since ethanol is one of the reactants in the biosynthesis of ethyl acetate, a significant correlation coefficient between their concentrations was expected. The calculated statistic (r^2) equals 0.810 (df = 19) which is significant at the 0.01 level.

The compound represented by the last column in Table II was only recently identified as acetoin. Time constraints and the lack of a pure standard with which to construct a standard curve precluded the presentation of the data in terms of molarity. Instead, the peak areas were made relative by dividing by the approximate average of agria, organ pipe, and senita peaks. This statistic, although somewhat contrived, clearly demonstrates that acetoin is much more concentrated in the saguaro and *Opuntia* plates. In fact, it appeared that acetoin was the predominant volatile produced by the action of *Erwinia* on these two cacti.

A two-level nested analysis of variance was performed on the data for each volatile except methanol. The resultant probabilities are shown at the bottom of Table II. Ethyl acetate was the only compound for which differences between cacti were not statistically significant. Differences between the amounts of ethanol and ethyl acetate produced by the two bacterial strains were marginally significant ($P \leq 0.1$).

The data in Table II and Vacek's data on the concentrations of volatiles observed in naturally occurring cactus rots (Table III) provide for several interesting comparisons. First, the two data sets are similar in that methanol concentrations in different cacti are fairly equal. Another similarity is that ethanol was not found in high concentration in any of the saguaro necroses. However, differences between the data sets outnumber the similarities. First, ethanol was found in much higher concentrations in field rots than in the lab experiments. Second, senita rots in nature, like saguaro, do not contain many volatiles. Third, ethyl acetate and acetoin are not listed as volatiles in naturally occurring rots while acetone, 1-propanol, and 2-propanol were not observed in the *Erwinia* rotted cacti. Vacek (1979)

TABLE III.   Average and Maximum Concentrations (mM) of
Five Volatiles in Naturally Occurring Cacti
Necroses

Volatile	Agria[a] N=17		Organ pipe[a] N=23		Saguaro[b] N=4		Senita[b] N=4	
	Mean	Max	Mean	Max	Mean	Max	Mean	Max
Methanol	2.8	6.0	3.2	9.5	5.0	7.5	5.1	5.9
Ethanol	13.6	55.8	7.3	33.9	0.2	0.7	0.3	0.5
Acetone	8.9	114.8	1.6	13.3	0.2	0.4	0.4	0.5
2-Propanol	22.3	109.8	5.3	31.3	0.0	0.0	0.2	0.4
1-Propanol	5.1	22.0	2.1	11.2	0.4	1.1	0.3	0.7

[a]data from Vacek (1979); samples associated with at least
one Drosophila life stage.
[b]data from Vacek published in Heed (1978;

stated that other unidentified volatiles were present in
field rots, but were not included in his analysis. Presumably, ethyl acetate and acetoin fall into this catagory.
With respect to the high concentration of acetone and the
propanol isomers in agria and organ pipe rots, it is obvious
that Er. carnegieana is not the sole manufacturer of volatiles in natural rots. The microbial communities in these
necroses appear to be quite diverse. Perhaps, as Young
et al. (1981) suggest, the obligate anaerobes are responsible for the production of acetone and the propanols. At
any rate, volatiles are certainly present in cactus rots in
significant concentrations, and yeasts and bacteria are both
apparently involved in their formation.

III   Volatiles and Larval Behavior

Investigations into the behavior of Drosophila larvae
have been rare. Larvae are known to have cuticular chemoreceptors as well as several larger sense cones (also
chemoreceptors) in the intima of the pharynx (Bodenstein,
1965). Using D. melanogaster larvae, Pruzan and Bush (1977)
demonstrated that larvae have the capacity for olfactory
discrimination. Larvae of certain strains of D.
melanogaster exhibited a strong preference for agar containing ethanol (6%), whereas larvae from strains derived from
sympatric populations of a sibling species, D. simulans, did
not show this preference (Parsons, 1977). Parsons (1979)
extended these studies to include three additional

fermentation products, acetic acid (0.4%), ethyl acetate
(0.3%), and lactic acid (2.0%), and one additional species,
D. immigrans. All three species had significant preferences
for ethyl acetate and the acids, and these preferences ex-
ceeded the preference for ethanol. In general, differences
in preference between geographic strains of a species were
not as great as differences between species.

The relevance of larval preference for particular compounds stems from the observation that larvae of some species
selectively feed on certain yeasts in their natural substrates (Fogleman *et al.*, 1981 and 1982). Of the cactophilic
Drosophila in the Sonoran Desert, *D. mojavensis* larvae were
shown to typically contain significantly higher percentages
of some yeasts (particularly *Pichia cactophila*) than their
substrates while *D. nigrospiracula*, *D. mettleri*, and *D.
pachea* generally contained random samples of available
yeasts. Three other species, *D. arizonensis*, *D.
pseudoobscura*, and *D. melanogaster*, which utilize necrotic
oranges as larval substrates in the Tucson area were also
found to be selective yeast feeders. Results of laboratory
tests support larval preference behavior rather than differential digestion as being primarily responsible. Larvae are
apparently capable of distinguishing between patches of
different yeast species in natural substrates and spend more
time feeding in patches of preferred yeasts. Volatile compounds are reasonable candidates for the chemical basis of
these preferences.

Larval preference for volatiles were determined in Petri
dishes of water-agar (2%) using *D. mojavensis* second instar
larvae. The volatiles were made up in water, typically at a
concentration of 10 mM in a suspension (215 mg/ml) of solvent
extracted yeast. One drop of this mixture was placed at 12
and 6 o'clock on the plates while control drops (yeast suspension without volatiles) were placed at 3 and 9 o'clock.
The dead yeast served to retain the larvae since, although
not an attractant, they will feed on it once there. The
drops were allowed to dry for about 30 minutes before 25
larvae were introduced into the center of each plate. Because
volatiles by definition evaporate easily, the actual concentrations to which the larvae were ultimately exposed are
unknown but certainly less than the initial concentrations.
The number of larvae in each of the 4 areas was recorded at
5, 10, 20, and 30 minutes after introduction. The data are
expressed in Table IV as the percent of larvae in the areas
with volatiles averaged over replications.

TABLE IV. Preference of D. mojavensis Larvae for
Volatiles. Preference is Indicated when the
Average Percent Observations of Larvae in the
Experimental Areas is Significantly Greater
than 50%

Volatile	Conc. (mM)	Reps.	Total obs.	Avg. % obs. in exp. areas ± S.D.	t
Methanol	10	4	216	45.4 ± 12.1	0.760
Ethanol	1	8	402	47.7 ± 8.1	0.803
	5	7	401	60.5 ± 8.4	3.307[a]
	10	4	131	59.0 ± 7.8	2.308
Acetone	5	4	323	55.0 ± 10.5	0.952
	10	4	267	45.4 ± 4.0	2.290
	20	8	394	61.4 ± 14.8	2.179
	40	8	339	54.8 ± 12.9	1.051
Ethyl acetate	1	8	392	52.1 ± 8.9	0.670
	5	8	438	60.7 ± 16.9	1.792
	10	4	337	71.9 ± 6.1	7.125[b]
1-Propanol	10	8	231	44.7 ± 16.0	0.940
2-Propanol	10	4	163	53.2 ± 14.7	0.435
1-Butanol	10	8	393	52.4 ± 17.1	0.398
iso-Butanol	10	4	203	60.8 ± 18.0	1.198
1-Pentanol	10	4	199	69.5 ± 14.4	2.716
iso-Pentanol	10	8	270	49.9 ± 20.1	0.014
active-amyl alc.	10	8	365	50.2 ± 12.3	0.046
neo-Pentanol	10	8	492	52.7 ± 11.2	0.679
2-Pentanol	10	8	339	51.3 ± 11.8	0.311
3-Methyl-2-BuOH	10	4	176	58.7 ± 9.7	1.790
3-Pentanol	10	8	366	58.4 ± 15.2	1.558
tert-Pentanol	10	4	206	50.0 ± 3.6	0.000

[a] $p<0.05$
[b] $p<0.01$

As can be seen in Table IV, statistically significant preferences were exhibited for only two volatiles, ethanol and ethyl acetate. Unfortunately, this type of experiment yields data that show high variance between replicates. Still, the results are in perfect agreement with those of Parsons' previously mentioned experiments on D. melanogaster. Ethanol and especially ethyl acetate are larval attractants. Referring back to Table I, ethyl acetate is made by three of the 13 cactophilic yeast species, and only two of these 3 are common in nature, P. cactophila and P. amethionina (Fogleman et al., 1981).

Additional preference tests were set up using yeasts
instead of volatiles. The same procedure was used except
that suspensions of live yeasts were dropped onto YM plates
(supplemented with 1% dry agria powder) in four areas and
allowed to grow for 24 hrs at 37°C before being used in the
tests. Second instar *D. nigrospiracula* and *D. mettleri*
larvae were then introduced into the plates as before. The
results are given in Table V.

The larval preferences for *D. nigrospiracula* and *D.
mettleri* were identical in that, of the six possible pair-
wise comparisons of the four yeast species, there was only
one in which the larvae did not express a significant preference. Larvae of both species were attracted to *P.
cactophila* and *P. amethionina* (v.p.) equally. Both of these
yeast species were significantly preferred over the other
two, *P. heedii* and *P. opuntiae.* The preference hierarchy for
larvae of both *Drosophila* species then is *P.a.* (v.p.) =
P.c.>P.h.>P.o. The first two species in the hierarchy produce ethyl acetate, and low concentrations of ethanol were
detected in the *P. heedii* cultures whereas none was found in
cultures of *P. opuntiae* (Table I). Therefore, the data from
this experiment support the contention that ethyl acetate is
a primary and ethanol is a secondary larval attractant. A
similar experiment performed with *D. mojavensis* larvae gave

TABLE V.   *Preference of D. nigrospiracula and D. mettleri
Larvae for Four Species of Cactophilic Yeasts.
Numbers of Replications are 5 for D.
nigrospiracula and 3 for D. mettleri*

Yeast[a] species		D. nigrospiracula		D. mettleri	
Y-1	Y-2	Tot. obs.	Avg. % obs. in Y-1 areas ± SD	Tot. obs.	Avg. % obs. in Y-1 areas ± SD
Pc vs. Pap		209	41.3 ± 20.0	72	42.6 ± 4.9
Pc vs. Ph		263	66.4 ± 10.2[b]	72	69.2 ± 3.6[b]
Pc vs. Po		205	81.0 ± 4.2[d]	72	75.8 ± 4.7[b]
Pap vs. Ph		228	78.4 ± 10.1[c]	87	82.1 ± 6.1[b]
Pap vs. Po		229	84.2 ± 3.1[d]	94	80.7 ± 2.5[c]
Ph vs. Po		177	71.6 ± 15.6[b]	53	78.2 ± 6.5[b]

[a]*Yeast species (strain number) are: Pc = P. cactophila
(76-243A), Pap = P. amethionina v. pachycereana (76-401B-α),
Ph = P. heedii (76-356), Po = P. opuntiae (76-211).*

[b]*p<0.05*
[c]*p<0.01*
[d]*p<0.001*

a preference hierarchy of *P.c.*>*C. sonorensis*>*C. ingens* (W.T. Starmer, pers. comm.). These results can also be explained on the basis of the concentrations of ethyl acetate and ethanol produced by these yeasts (Table I).

The correlation between larval preference for yeast species and the volatiles that they produce can be used to explain the observations on larval selectivity in natural substrates (Fogleman *et al.*, 1981 and 1982). In comparisons of the yeast florae in natural substrates and larval guts of *D. mojavensis*, the only comparisons which did not indicate significant selective feeding on *P. cactophila* (2 out of 6) were those where *P. cactophila* and *P. amethionina* were the predominant yeasts (*P.c.*+*P.a.* = 100% in one case and 80% in the other). Again, these two yeasts are equally attractive. Selective feeding was not shown in larvae-substrate comparisons of *D. pachea* because *P. cactophila* and *P. amethionina* are not typical residents of senita rots (Starmer *et al.*, 1982). Perhaps larvae of *D. nigrospiracula* and *D. mettleri* do not selectively feed in nature because saguaro tissue does not support the production of ethanol or ethyl acetate by microorganisms. This is suggested by the absence of these two compounds in the volatile profiles produced by *Erwinia* (Table II).

All of the above results lead to the question of why larvae are attracted to ethyl acetate in the first place. Some data on cactophilic yeasts and *D. buzzatii* in Australia link larval preference to nutritional quality (Vacek, 1982). While there is no corresponding data for the Sonoran Desert *Drosophila*, it is certainly an attractive hypothesis. An alternative is that it is merely the larval manifestation of an adult preference which may be more ecologically important (see next section). Obviously, more research is needed to fully elucidate the ultimate basis of larval preference.

IV   Volatiles and Adult Behavior

One can ascribe several roles to volatile compounds in the ecology of adult *Drosophila*. First, they serve as general attractants to food sources or potential breeding sites. In the Sonoran Desert, differences between cactus species with respect to volatile profiles (Tables II and III) may be the basis of host plant selection. Once the flies have located a substrate, volatiles may also serve as oviposition stimulants. Finally, volatiles may stimulate mating in general or affect certain aspects of mating such as mate selection. Since little if any information exists on this last point, only the first two subjects will be reviewed.

A variety of chemicals have been tested as to their attractiveness to *Drosophila* adults. It is not suprising that most of those found to be significant attractants represent the common by-products of microbial metabolism. These include ethanol, ethyl acetate, amyl alcohol, acetic and lactic acids, acetaldehyde, diacetyl, acetoin, and indole (Barrows, 1907; Hutner *et al.*, 1937). Mixtures of attractive volatiles were much more attractive than single compounds, and certain mixtures approached natural substrates in attractiveness (Hutner *et al.*, 1937). Other generalizations which are supported in the literature are that larger flies and females are more responsive to odors than smaller flies and males, adults respond to a range of concentrations above which the volatile is repellent (Reed, 1938), and different *Drosophila* species respond differently (West, 1961).

The investigations discussed above leave little doubt that *Drosophila* locate feeding and breeding substrates by positive chemotaxis toward fermentation by-products. The olfactory sense organs which are responsible for this directed movement are located in the third or terminal segment of the antennae (Barrows, 1907). Antennaless phenotypes of *D. melanogaster* do not respond to attractive chemical stimuli (Begg and Hogben, 1946). Orientation towards the substrate is accomplished by adjusting for unequal stimulation of the antennae since removal of one antenna results in circular fly movement when exposed to food odors (Barrows, 1907).

The genetic basis of olfactory response has only recently been examined and remains obscure. Fuyama (1976) reported that differences between 5 geographical strains of *D. melanogaster* in their response to ethanol, lactic acid, ethyl acetate, and n-butyraldehyde were highly significant. While this suggests that genetic variation for olfactory response exists in natural populations, Fuyama appropriately pointed out that the observed inter-strain differences could have been due to behaviors irrelevant to olfaction such as general activity levels, geotaxism, or differential tolerance to starvation. Subsequently, a genetic basis for olfactory response to ethyl acetate in *D. melanogaster* was demonstrated using chromosomal lines extracted from natural populations (Fuyama, 1978). The right arm of the second chromosome appears to contain genes which determine a large fraction of the observed odorant-specific variation. The extension of this type of study to other species and to genetic systems that are specific for other odors is both pertinent and desirable.

Since there are interspecific differences with respect to chemoreceptivity, one may ask if the cactophilic *Drosophila* are also attracted to volatiles. In a field test using 5 species of cacti (agria, organ pipe, saguaro, senita and cina) artificially rotted with *Er. carnegieana*, Fellows and Heed (1972) demonstrated that the host plant discrimination for feeding adults of all 4 species was very high but not absolute. That is, each species was attracted to its normal host plant, but its specificity was reduced compared to data from natural rot pockets.

In order to implicate specific volatile compounds as attractants to the cactophilic *Drosophila*, several laboratory experiments were done. Mixtures of four volatiles, methanol, ethanol, methyl acetate, and ethyl acetate, were added to fresh agria, organ pipe, saguaro, and senita homogenates. The attractiveness of these mixtures to adult flies was compared to controls (homogenates without volatiles). The four volatiles were added to the homogenates at concentrations representing multiples of the data set (averaged over bacterial strains and replications) presented in Table II. Homogenates were put into shell vial traps and duplicate experimental and control traps (16 total) were positioned on a random grid in a small room (6' x 12'). Several thousand flies were then released *en masse*, and the number of flies in each trap after 24 hrs was recorded. The results of the first test showed that *D. nigrospiracula* was not significantly attracted to traps containing volatiles at 2X the average concentrations listed in Table II. Two additional tests showed that *D. mojavensis* was not attracted to 4X or 16X these concentrations either. However, when the original *Erwinia* rotted material (from which the data in Table II was collected) was used, a highly significant preference for traps containing volatiles was expressed by *D. mojavensis* (F=7.338; df=4,8; P<0.01). Differences between the attractiveness of the different cactus species were not statistically significant. Evidently, initial attraction of cactophilic *Drosophila* involves more than a mixture of these 4 compounds. The presence of the fermentation acids or acetoin, neither of which were added to the homogenates, could be of major importance to the general attractiveness of the cacti and the host plant specificity of the flies. Hutner *et al.* (1937) found that mixtures containing acetoin (as well as diacetyl, indol, and acetaldehyde) were much more attractive to *D. melanogaster* than the common fermentation

alcohols, acids, and esters. Again, a more complete understanding of the relationship between volatiles and adult
attraction demands further research.

Once the flies have arrived at the rot site, volatiles
may influence feeding and/or oviposition behavior. It is well
known that *Drosophila* adults exhibit feeding preference for
certain yeasts over others (Vacek, 1982 and references
therein). Like larvae, adults may use volatile compounds as
cues to distinguish between different yeast species. Several
tests of oviposition preference for medium containg 9% ethanol (2.2M) over medium containing no alcohol have shown the
existence of a large amount of interspecific variation in
this behavior with some species laying more than 90% of their
eggs on the alcohol medium and other species clearly avoiding
it (McKenzie and Parsons, 1972; Richmond and Gerking, 1979).
Finally, acetic acid,which is a known attractant,was recently
shown to be an oviposition stimulant (Fluegel, 1981).

V Epilogue

This paper represents an effort to describe the involvement of volatile compounds in the ecology of cactophilic
Drosophila. Towards this goal, it has only been partially
successful. Numerous facets and functions have yet to be
fully elucidated. Noteworthy among these are the role of
volatiles in host plant selection, the genetic systems which
control chemoreceptivity and response, and the relationship
between volatile by-products of microorganismic activity and
fitness parameters such as nutrition. What can be said with
some finality is that volatiles are fundamental, pervasive,
and salient factors in the ecology of the cactus-micro-
organism-*Drosophila* model system and their continued study
will undoubtably lead to significant contributions to the
undertanding of this system.

REFERENCES

Alcorn, S., and May, C. (1962). *Plant Dis. Rep. 46*, 156.
Barrows, W. (1907). *J. exp. Zool. 4*, 515.
Begg, M., and Hogben, L. (1946). *Proc. R. Soc., Ser. B133*, 1.
Bodenstein, D. (1965). *In* "Biology of Drosophila" (M.
    Demerec, ed.), p. 275. Hafner Publ. Co., New York.

Boyle, A. (1949). *Phytopathology 39*, 1029.

Collins, R., and Morgan, M. (1960). *Science 131*, 933.

Collins, R., and Kalnins, K. (1965). *Phyton 22*, 107.

Fellows, D., and Heed, W. (1972). *Ecology 53*, 850.

Fluegel, W. (1981). *J. Insect Physiol. 27*, 705.

Fogleman, J., Starmer, W., and Heed, W. (1981). *Proc. natn. Acad. Sci. U.S.A. 78*, 4435.

Fogleman, J., Starmer, W., and Heed, W. (1982). *Oecologia* (in press).

Fuyama, Y. (1976). *Behav. Genet. 6*, 407.

Fuyama, Y. (1978). *Behav. Genet. 8*, 399.

Graf, P. (1965). M.S. Thesis, Univ. of Arizona.

Gyosheva, B., and Rusev, P. (1979). *Die Nahrung 23*, 385.

Heed, W. (1978). *In* "Ecological Genetics: The Interface" (P.F. Brussard, ed.), p. 109. Springer-Verlag, New York.

Hutner, S., Kaplan, H., and Enzmann, E. (1937). *Am. Nat. 71*, 575.

Ingraham, J., and Guymon, J. (1960). *Archs. Biochem. Biophys. 88*, 157.

Lightle, P., Standring, E., and Brown, J. (1941). *Phytopathology 39*, 1029.

McKenzie, J., and Parsons, P. (1972). *Oecologia 10*, 373.

Parsons, P. (1977). *Oecologia 30*, 141.

Parsons, P. (1979). *Aust. J. Zool. 27*, 413.

Phaff, H., Miller, M., and Mrak, E. (1978). "The Life of Yeasts." Harvard Univ. Press, Cambridge, Mass.

Prior, B., Kilian, S., and Lategan, P. (1980). *Arch. Microbiol. 125*, 133.

Pruzan, A., and Bush, G. (1977). *Behav. Genet. 7*, 457.

Reed, M. (1938). *Physiol. Zool. 11*, 317.

Richmond, R., and Gerking, J. (1979). *Behav. Genet. 9*, 233.

Schink, B., and Zeikus, J. (1980). *Cur. Microbiol. 4*, 387.

Starmer, W. (1982). *Microbial Ecology (in press).*

Starmer, W., Phaff, H., Heed, W., Miranda, M., and Miller, M. (1982). *Evol. Biol. 14*, 269.

Steenbergh, W. (1970). *J. Ariz. Acad. Sci. 6*, 78.

Tabachnik, M., and DeVay, J. (1980). *Physiol. Plant Path. 16*, 109.

Thayer, D., and Ogg, J. (1967). *J. Bact. 94*, 488.

Vacek, D. (1979). Ph.D. Dissertation, Univ. of Arizona.

Vacek, D.C. (1982). Chapter 12, this Volume.

VanVuure, H., Toerien, D., and Lategan, P. (1978). *S. Afr. J. Sci. 74*, 387.

West, A. (1961). *J. Econ. Entomol. 54*, 677.

White, J., and Starr, M. (1971). *J. Appl. Bact. 34*, 459.

Young, D., Vacek, D., and Heed, W. (1981). *Drosophila Inf. Serv. 56*, 165.

PART IV
POPULATION GENETICS
AND ECOLOGY

14
Population Genetics of *Opuntia* Breeding *Drosophila* in Australia[1]

J. S. F. Barker

Department of Animal Science
University of New England
Armidale, New South Wales

Studies of the cactus-yeast-*Drosophila* system in
Australia have focussed primarily on the *Drosophila* and on
the mechanisms maintaining genetic variation at allozyme loci.
There is now a large body of data on gene frequencies and
heterozygosities for a variety of allozyme loci in a multi-
tude of species. Analyses of these data, attempting to relate gene frequency distributions to expectations of various
theoretical models continue to provide equivocal results, depending on the assumptions of particular models, and real
doubts about the validity of pooling results from different
allozymes and many different species. This statistical mani-
pulation of data from many loci and species cannot contribute
definitively to solution of the questions of interest, *viz.*
at what proportion of allozyme loci is genetic variation
actively maintained by some form of selection, how do the relative fitnesses of genotypes relate to ecological and environmental variation, and to the biochemical properties and
function of the enzyme gene products, and thus what mechanisms operate to maintain allozyme polymorphism in natural
populations.

Sound empirical studies of suitably chosen species in
both natural and laboratory populations are needed. The cactophilic *Drosophila* are near to ideal (Barker, 1977; Heed,
1978), primarily because their breeding and feeding site is
known, so that effects of biotic environmental factors, *viz.*
the microorganisms on which the *Drosophila* may feed and the

[1]Supported by the Australian Research Grants Scheme

ECOLOGICAL GENETICS AND EVOLUTION
ISBN 0 12 078820 9

cactus which provides the substrate for both yeasts and
Drosophila, can be investigated. The cactophilic *Drosophila*
are members of the *mulleri* subgroup of the *repleta* group
(Wasserman, 1982). All are native to the Americas, but two
have colonized in Australia, *viz. D. buzzatii* (Mather, 1957)
and *D. aldrichi* (Mulley and Barker, 1977). Both species are
commonly found breeding in the same rot in areas of Australia
where their distributions overlap. However, differential
yeast preferences of these species (Barker *et al.*, 1981a, b)
indicate a degree of niche separation, which may be due
to separate *Drosophila*-yeast evolution in the different
regions of origin, or which may still be evolving in
Australia.

Comparative studies of *D. buzzatii* and *D. aldrichi* form
an important part of our experimental program, and the opportunity for study of competition and niche separation in these
newly sympatric populations is an exciting prospect. Such
studies would complement those on the endemic *Drosophila*
species of the Sonoran Desert (Heed, 1978). The cactus-yeast-
Drosophila system promises to be of major import for studies
in evolutionary biology.

I   Distribution and Population Structure of *D. buzzattii* and *D. aldrichi*

The establishment, spread and control of *Opuntia* cactus
in Australia are described by Barker and Mulley (1976) and
Murray (1982). The present *Opuntia* distribution is essentially within the limits of the original infestation (Fig. 1),
but occurring as an island distribution with island size
ranging from just a few plants to a few hundred hectares. In
southern N.S.W., Victoria and South Australia, outside the
limits of the original infestation, some patches of pest pear
also remain, but most islands comprise from one to about 30
plants of the cultivated species, *O. ficus-indica*.

Available evidence (Barker and Mulley, 1976) indicates
that *D. buzzatii* is specific to the cactus niche. *D. aldrichi*
is assumed to be similarly restricted. The population structure of both species thus depends primarily on the distribution of *Opuntia*, and secondarily on environmental factors
that may limit their distribution and abundance. *D. buzzatii*
appears to exist throughout the entire *Opuntia* distribution,
but *D. aldrichi* has been found only in Queensland, with one
exception (locality 5, Fig. 1), separated by some 900 km from
the southern end of its main distribution.

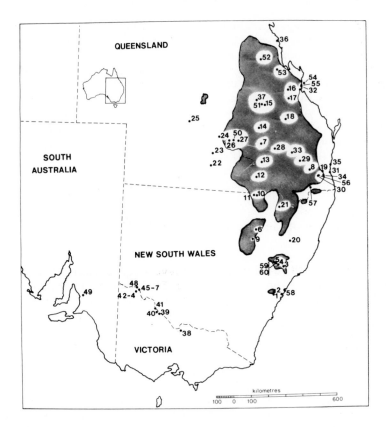

FIGURE 1. Localities from which D. buzzatii populations have been sampled, and the distribution of the main Opuntia infestations in 1920 (shaded areas).

As *D. buzzatii* and *D. aldrichi* have not been raised from fruit of *Opuntia* species (except for *D. buzzatii* from *O. ficus-indica* in southern Australia, and once from *O. stricta*), maintenance of populations of both species apparently is dependent on cladode rots. In *O. stricta* and young plants of the tree pears, the majority of rots develop from bacterial infections that follow damage to the plants due to larval *Cactoblastis cactorum* (Murray, 1982) feeding voraciously on the plant tissue within stems or cladodes. In addition, some rots result from infestation of the plant by the cochineal scale insects (*Dactylopius* spp. Costa) or from physical injury. Mature tree pears are rarely attacked by *C. cactorum* and rots are due to cochineal scale insects, physical damage or tissue degeneration in old cladodes. As the cochineal insects were first established about the same time as *C. cactorum* was being extensively distributed, the cactophilic *Drosophila* could not have become established prior to this time. Thus we have argued (Barker and Mulley, 1976) that both species entered Australia in material imported for the biological control program, probably about 1925–1930. At this time, they would have had an almost unlimited environment, and must have produced enormous populations and spread rapidly through the *Opuntia* infestation. As the pear was controlled and its distribution receded by 1940 to the islands as found today, so the cactophilic *Drosophila* would have contracted to spatially isolated populations.

II Maintenance of Genetic Variation in *D. buzzattii*

A total of 36 allozyme loci have been assayed for a maximum of 99 population samples from 56 localities (Barker, 1981). Of these loci, eight are polymorphic, *viz. Esterase-1 (Est-1), Esterase-2 (Est-2), β-N-acetyl-hexosaminidase (Hex), Phosphoglucomutase (Pgm), Aldehyde oxidase (Aldox), Alcohol dehydrogenase-1 (Adh-1), Alcohol dehydrogenase-2 (Adh-2)* and *Leucine aminopeptidase (Lap)*. Average heterozygosity (0.051 ± 0.025) is lower than that for most other species of *Drosophila*, but average heterozygosities based on eight loci common to our work and that of Zouros (1973) for other cactophilic species of the *mulleri* subgroup were not significantly different (Barker, 1977). Low levels of genetic variability may be characteristic of the cactophilic *Drosophila*.

While selection may not be acting directly on the allozyme loci, results of a spatial study of gene frequencies indicated that some form of balancing selection was operating (Barker and Mulley, 1976). Further, genetic drift and migration as a sufficient explanation for observed genotype-environment associations was excluded by Mulley *et al.* (1979). Their results suggested that some form of selection has been operating (at least for *Est-2, Adh-1* and *Pgm*) and that there has been differential adaptation of the different populations to the different climates to which they have been exposed over the last 40–50 years. More direct evidence (for *Est-2, Adh-1* and *Hex*) has been obtained from a perturbation experiment in one natural population (Barker and East, 1980). For *Adh-1*, significant gene frequency differences between coastal and inland populations have been interpreted as due to selection (Barker, 1977), while gene frequency and heterozygosity data of flies emerging from individual rots imply differential selection between the sexes for *Adh-1* and a possible interaction between this locus and *Aldox* (Barker, 1981).

Assuming then that at least some allozyme polymorphisms are maintained by selection, the question remains as to its nature. Heterosis as an explanation would seem quite untenable (Lewontin *et al.*, 1978), but not necessarily for all selection models (Karlin, 1981). However, environmental heterogeneity allows the possibility of diversifying selection (Hedrick *et al.*, 1976) or habitat choice (Jones, 1980; Powell and Taylor, 1979), both of which are plausible mechanisms. For the cactophilic *Drosophila*, our working hypothesis is that heterogeneity among and within rots in (a) microflora composition and (b) chemical composition, itself partly dependent on (a), could be important. Much of our work is directed towards determining the effects of these and other microenvironmental factors, and their heterogeneity, on relative fitnesses of allozyme genotypes and on heterozygosity. Studies of the yeasts are reviewed by Vacek (1982) and some aspects relating to chemical composition are discussed below, and by East (1982) in relation to esterases.

III   Identification of Putative Selective Factors

The available evidence provides strong support for the hypothesis that selection is influencing allozyme gene frequencies at some loci. However, if the nature of this selection is to be understood, the environmental factors involved need to be identified and related to the enzyme products and their function. The significant genotype-environment associations (Mulley *et al.*, 1979), based on genetic data from many populations and climatological variables, may not be due to direct effects of the environmental variables. Further, the climatological data may not reflect the environment as experienced by the *Drosophila*.

Thus we have sought (a) associations between genetic variables and environmental variables measured over time within a single population (temporal study) and (b) associations between genetic variables in flies emerging from individual rots and environmental variables measured for each rot. For the latter, two sets of data have been obtained; one on 18 rots during Collection 45 of the temporal study, the other on 100 rots from four localities in southern Queensland (coastal and inland rots).

A. *Genotype-Environment Associations Within One Population*
   (Temporal Study)

One population of *D. buzzatii* (locality 5, Fig. 1) was
sampled essentially every month for four years (46 samples)
and collected flies assayed for the six loci polymorphic in
this population. As the number of flies collected varied
widely among months (from 9 to 1648), standardized gene frequency deviations (Christiansen *et al.*, 1976) were used in
the analyses rather than estimated gene frequencies. Canonical correlation analyses were done for each of standardized
gene frequency deviations, expected and observed heterozygosities with the environmental variables COLLUNIT to H and
MAXTEMP to ROTINC (as listed in Table I), SD's and CV's. The
only significant correlation was for standardized gene frequency deviations with COLLUNIT to H (R = 0.76, P = 0.012),
while that for observed heterozygosities with SD's was near
to significance (R = 0.72, P = 0.08). Multiple regression
analyses of each genetic variable on these sets of the environmental variables showed significant associations with one
or more environmental variables for gene frequencies at each
locus and for expected and observed heterozygosities (except
Est-1 and *Est-2*). The two multivariate analyses did not
always both detect a particular association as significant,
but when an association was significant in one analysis, it
was generally consistent (i.e., in the same direction) in the
other analysis.

Gene frequencies at every locus (except *Adh-1*) showed
significant associations with *D. buzzatii* population size
variables (COLLUNIT, NOBUZZ or POPSIZE). Such associations
may indicate density-dependent selection, but equally they
could be due to other environmental variables that affect
gene frequencies and also are reflected in the changes in
D. buzzatii population size. Similarly, significant associations of gene frequencies with NOALD may indicate effects of
competition by *D. aldrichi*, or effects of environmental variables that are indicated by changes in the population size
of *D. aldrichi*.

Variability in temperature and relative humidity over the
28 days preceding each collection was associated with gene
frequencies at *Est-2* and *Aldox* and with *Pgm* and *Adh-1* observed heterozygosity, but the last also showed significant
associations with MAXTEMP, MINTEMP and RANGETEMP, i.e., seasonal variation, being high in summer and low in winter. A
possible selective effect on heterozygosity *per se* at the
Adh-1 locus is indicated, but it is not known whether this is
due to seasonal changes in environmental variability, in temperature or in some other correlated factor.

TABLE I.   *Environmental Variables Defined for Multi-*
variate Analyses of Data from the Temporal
Study

COLLUNIT	- *Number of collecting units, higher numbers indi-cating more time and effort expended and fewer flies in the population*
NOBUZZ	- *Number of* D. buzzatii *collected*
NOALD	- *Number of* D. aldrichi *collected*
POPSIZE	- *NOBUZZ/COLLUNIT*
RATIO	- *NOBUZZ/NOALD*
H	- *Shannon-Weaver species diversity index*
MAXTEMP[a]	- *Mean daily maximum temperature*
MINTEMP	- *Mean daily minimum temperature*
RANGETEMP	- *Mean daily temperature range*
EXTMAX	- *Highest maximum temperature*
EXTMIN	- *Lowest minimum temperature*
EXTRANGE	- *(EXTMAX-EXTMIN)*
RAIN	- *Total rainfall*
RAINDAY	- *Number of days on which rain fell*
RELHUM	- *Mean daily 1500 hours relative humidity*
ROTINC	- *Subjective scoring of incidence of rots in the study location (coded 1, 2, 3)*
MAXSD, MINSD, RANGESD, RELHUMSD	- *Standard deviations of daily MAXTEMP, MINTEMP, RANGETEMP and RELHUM*
MAXCV, MINCV, RANGECV, RELHUMCV	- *Coefficients of variation of same four variables*

[a]*All weather variables are for the 28 day period prior to each collection.   EXTMAX, EXTMIN, EXTRANGE measured at the study location from Collection 13; previous to this and all other weather variables taken from records of the nearest Australian Bureau of Meteorology station (located 29 km away).*

B. *Genotype-Environment Associations (Collection 45)*

During Collection 45 of the temporal study (October, 1977), 18 rots were found where *Drosophila* larvae were present. For each, pH and temperature within the rot and ambient temperature were recorded, and the rot returned to the laboratory. The next day samples of each rot (pooled from multiple positions) were streaked onto nutrient agar plates (Wagner, 1944) and incubated at 25°C. All cultures were identified to species by Professor H.J. Phaff and Ms M. Miranda. *Drosophila* emerging from each rot were collected, and the *D. buzzatii* assayed for the six polymorphic allozyme

loci. No flies emerged from one rot, *D. buzzatii* from 17
(15 to 672 flies) and *D. aldrichi* from four (1 to 31 flies).
The environmental variables defined for each rot for
multivariate analyses are given in Table II. Preliminary
analyses of these data (done before the yeast species were
identified) were discussed briefly by Barker (1981). In
total, seven species of yeasts were isolated from the 17 rots
included in the analyses, but the analyses used only the presence or absence of the three most common species (*viz.*
Candida sonorensis, Pichia cactophila and *Clavispora* species
"O". Canonical correlation analyses were done separately
for standardized gene frequency deviations with NOBUZZ
to ROTAGE (excluding ROTTEMP, which was highly correlated
with TEMPDEV, r = 0.83, P < 0.001) and with NO YSP to *Cl.o.*
Only the former canonical correlation was significant (R =
0.99, P = 0.005). Canonical correlation analyses of observed
and expected heterozygosity with each of the above two sets
of environmental variables and with all included also were
done; only that for expected heterozygosity with all environmental variables was significant (R = 1.0, P < 0.001).
These analyses indicate that the allele *Est-1^c* increases
in frequency, while *Est-2^b* and *Adh-1^b* decrease with increasing age and pH of the rot and increasing number of *D.buzzatii*
emerging from the rot. Further, the expected heterozygosity
for *Hex* increases with shorter rot duration and fewer yeast
species per rot, but with *Clavispora* sp. "O" present. All
these relationships are either supported, or at least not
contradicted by multiple regression analyses.

C. Genotype-Environment Associations (Coastal and Inland Rots)

Some preliminary analyses of this experiment were given
by Barker (1981), where it was noted that the primary purpose
was to test a specific hypothesis that the higher frequencies
of *Adh-1^b* in coastal populations could be some function of
higher concentrations of sodium chloride in this environment.
That hypothesis was not supported by the analyses, although
it was pointed out that the effect may be indirect and
mediated through the yeast species present in the rots. Data
on the yeast species were not available at that time.
The suggestion that *Adh-1* gene frequencies could be
affected by salt in the environment derived from laboratory
population cages of *D. melanogaster*, where *Adh^F* frequency
increased with increasing salt concentration (Barker, 1981).
Such an effect is unique, and no biochemical explanation is
obvious, but it is not restricted to *Adh* in *D. melanogaster*.

TABLE II. *Environmental Variables Defined for Each Rot*
          for Multivariate Analyses of Data from
          Collection 45

NOBUZZ	- *Number of D. buzzatii emerging*
ALD	- *Presence (1) or absence (0) of D. aldrichi among* *emergences*
ROTTEMP	- *Temperature in the rot*
TEMPDEV	- *(ROTTEMP - Ambient temp.)/Ambient temp.*
pH	- *pH of the rot*
ROTDUR	- *Subjective scale (0, 1) estimate of the length of* *time the particular cladode had been rotting (rot* *duration)*
ROTAGE	- *Subjective scale (0, 1) estimate of the age of that* *part of the cladode that was still rotting*
NO YSP	- *Number of yeast species isolated from the rot*
C.s.	- *Presence (1) or absence (0) of C. sonorensis*
P.c.	- *Presence (1) or absence (0) of P. cactophila*
Cl.o	- *Presence (1) or absence (0) of Clavispora sp. "0"*

Two laboratory populations of *D. buzzatii* maintained for 29
months on a medium containing 0.75% NaCl have shown a steady
significant *decrease* in $Adh-1^b$ frequency (Fig. 2). This also
does not support the hypothesis, but the results for
D. melanogaster and *D. buzzatii* are consistent, both showing
an increase in the frequency of the higher activity allele
(Oakeshott *et al.*, 1982). The higher frequency of $Adh-1^b$ in
coastal populations remains an enigma, and is further complicated by gene frequencies in isofemale lines from Brasil and
Argentina. These were all collected from inland populations,
yet overall gene frequencies of $Adh-1^b$ were 0.95 in Argentina
and 0.86 in Brasil.
      In the field experiment in southern Queensland, 50 rots
were assayed from each of a coastal population (Hemmant -
locality 31) and the nearest inland populations (Grandchester,
Grandchester Hill and Borallon - localities 30, 56, 19; 50-65
km inland). For each rot, temperature and pH were measured
at the time of sampling, and variables relating to age of the
rot and type of cladode were recorded. Samples of the rot
were taken for assay of the concentrations of yeast species
present, and where possible (i.e., sufficient rotting material)
for analyses of Cl^-, Na^+, K^+, Mg^{2+} and Ca^{2+} ion concentrations, and of alcohol (methyl alcohol, ethyl alcohol, iso-
propanol) and acetone concentrations. As sufficient rotting
material was not available for many rots, samples of living
cladode tissue were taken adjacent to each rot for analysis

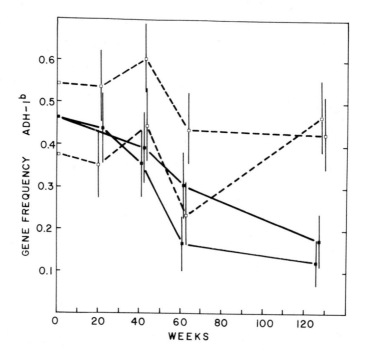

FIGURE 2. $Adh-1^b$ *gene frequencies in two replicate*
population cages of D. buzzatii, maintained on medium con-
taining 0.75% sodium chloride (■) and in two control popula-
tions (□).

of ion concentrations. The rots were returned to the laboratory and emerging flies collected and assayed for known polymorphic loci, 75 rots producing sufficient *D. buzzatii* adults
for gene frequency estimation. Five sets of environmental
variables were defined:

1. *Rot Variables*. ROTTEMP, TEMPDEV, pH, ROTDUR, ROTAGE
and NOBUZZ as for the Collection 45 rots. In addition, the
following were defined: (a) CLADODE TYPE – Young cladode
(coded 1) and old cladode (coded 2) – aerial cladodes, the
former current season growth, the latter previous season or
older, and basal cladodes (coded 3) – ones lying on the
ground, generally old and rooted, (b) NOALD – Number of emergences of *D. aldrichi*, (c) NO. SPECIES – Number of species of
Drosophila emerging from the rot, (d) NO. OTHER FLIES – Number of adult flies of species other than *D. buzzatii* and
D. aldrichi emerging from the rot.

2. *Yeast Variables.* (*a*) NO YSP - Number of yeast
species isolated from the rot, (*b*) Log_{10} concentrations of
each of the most common yeast species - *Candida sonorensis,
Pichia cactophila, Clavispora* species "O", *Cryptococcus
cereanus, Candida mucilagina* and *Candida boidinii.*

3. *Cladode Ion Concentrations* (CLCL, CLNA, CLK, CLMG,
CLCA).

4. *Rot Ion Concentrations* (ROTCL, ROTNA, *etc.*).

5. *Rot Alcohol Concentrations* (MEOH, ETOH, ISOPROP,
ACETONE).

Sets 1 and 2 were available for all 75 rots with sufficient *D. buzzatii*, while set 3 was available for 71 of these,
and sets 4 and 5 for only 22 of the rots.
In preliminary multivariate analyses (Barker, 1981),
estimated gene frequencies were used, but because of the
varying numbers of flies assayed per rot, standardized gene
frequency deviations have been used here, calculated as deviations from the overall frequencies in each locality. As
this largely corrects for locality, multiple regressions were
done using gene frequencies, and fitting locality (coded 1-4)
first, then sets of environmental variables. Localities were
significantly different for *Est-2, Adh-1, Lap* and *Aldox* gene
frequencies, with all but *Aldox* due to differences between
the coastal and the inland populations.
However, as our primary concern was to detect any associations between sets of genetic variables and sets of environmental variables for an array of individual rots, standardized gene frequency deviations were used for canonical correlation analyses. Each of these deviations, observed and
expected heterozygosities and the environmental variables
were first expressed as residuals from regression on locality.
Thus the canonical correlations were between sets of variables after correction for differences among localities, and
were done for each of standardized gene frequency deviations,
observed and expected heterozygosities with each set of environmental variables and with sets 1 and 2 combined, 1, 2
and 3 combined and 4 and 5 combined.
For the rot and yeast variables, either separately or
combined, no canonical correlations were significant,
although there were high loadings for *Est-2* and *Adh-1* gene
frequencies and expected *Adh-1* heterozygosity with ROTDUR,
NO YSP, *Cl.o.* and *C.m.* When the cladode ion concentrations
were added to the environmental variables, similar

associations were detected, together with a high loading for potassium (CLK), giving a canonical correlation for the gene frequencies of 0.86 (P = 0.077). For the cladode ion concentrations only with gene frequencies (R = 0.62, P = 0.074), the main association was primarily decreases in $Est\text{-}2^a$ and $Est\text{-}2^c$ with decreasing calcium (CLCA) and increasing magnesium (CLMG). For the rot ion and alcohol concentrations, data were available only for 22 rots, but the canonical correlation between standardized gene frequency deviations and alcohol concentrations was significant (R = 0.96, P = 0.015). The correlation with the rot ion concentrations was not significant, and although near to significant when both rot ion and alcohol concentrations were included (R = 0.99, P = 0.077), the loadings were primarily on the alcohols.

Results for the multivariate regression analyses generally were consistent with the canonical correlation analyses. There were few significant and/or consistent associations of any of the rot variables with genetic variables. However, there was a variety of consistent associations between gene frequencies and heterozygosities at all loci (except *Pgm*) with yeast species in the rot and/or chemical composition of the plant or the rot tissue itself.

IV   Genotype-Environment Associations—Conclusions

It should be emphasised that the aim of these experiments and analyses was to identify possible selective factors, and thus to provide a basis for detailed analytical studies in both the field and laboratory.

In order to assess the relative importance of the different sets of environmental variables, the proportions of possible associations that were significant for each environmental set and locus have been summarized (Table III). The organics and the *Drosophila* community are most important for allele frequencies, with the former also most important for observed heterozygosity. For all loci except *Est-2* and *Pgm*, organics are most important for allele frequencies, although *Hex* also shows a high proportion of significant associations with the *Drosophila* community and *Adh-1* with yeasts. The proportion of significant associations is very much less for observed heterozygosities than for allele frequencies, although the organics again show high proportions for *Hex* and *Aldox*. In most cases (particularly for *Est-1* and *Est-2*), significant associations for allele frequencies are not reflected in associations for observed heterozygosity,

TABLE III. *Proportions of Possible Associations that were Significant in Canonical Correlation and Multiple Regression Analyses of the Temporal Study and Coastal and Inland Rots for (a) Alleles, and (b) Observed Heterozygosity*

	Drosophila community[a]	Macro climate[b]	Cactus microenvironment				
			Cladode ions[c]	Rot ions[d]	Organics[e]	Yeasts[f]	Overall
(a) Est-1	.250	0	0	.250	.500	0	.106
Est-2	.167	.037	.367	.100	.167	.167	.130
Hex	.583	.056	0	0	.625	0	.156
Pgm	.333	.042	0	0	0	0	.080
Aldox	.083	.056	0	0	.500	0	.078
Adh-1	.083	.028	.100	0	.500	.500	.156
Lap	—	—	.050	.100	.375	0	.107
Overall	.242	.033	.118	.091	.352	.091	.116
(b) Est-1	0	0	0	0	0	0	0
Est-2	0	0	0	0	0	0	0
Hex	.083	.083	0	0	.750	.071	.122
Pgm	0	.056	0	0	0	0	.022
Aldox	.083	.083	0	0	.625	0	.100
Adh-1	0	.194	0	.300	0	.286	.156
Lap	—	—	.100	0	.125	0	.048
Overall	.028	.069	.014	.043	.214	.051	.065

[a] COLLUNIT to H (Table 1) [b] MAXTEMP to RELHUMCV (Table 1) [c] CLCL to CLCA
[d] ROTCL to ROTCA [e] MEOH to ACETONE [f] NO YSP to C.b.

suggesting that any selection is not favouring heterozygotes
per se. On the other hand, *Adh-1* observed heterozygosity
shows high proportions of significant associations with rot
ions and macroclimate, but no or few associations of allele
frequencies with these environmental variables. Clearly
these associations provide presumptive evidence for selection,
although it is recognized that any selection may not be
acting directly on the allozyme loci and that a significant
association with a particular environmental variable does not
prove that that variable is itself influencing fitness differences among genotypes. However, the summary of the significant associations emphasises the potential importance of
microenvironmental factors affecting gene and genotype frequencies at allozyme loci, and the need for their detailed
study.

   In the temporal study, gene frequencies at every locus
except *Adh-1* showed significant associations with measures of
D. buzzatii population size. Although the seasonal peaks in
population abundance tend to coincide with the times of major
C. cactorum larval activity and therefore rot abundance, one
might expect higher larval density per rot when population
size is high. Thus associations with the number of
D. buzzatii emerging per rot would be predicted. For Collection 45, where rots were chosen for assay because larvae were
present in substantial numbers, such associations were found
for *Est-1*, *Est-2* and *Adh-1*, but NOBUZZ showed no significant
associations with gene frequencies in the coastal and inland
rots. However, comparison of the emergence records for the
two sets of rots indicate that in the latter the flies collected were in many cases from the end of an emergence distribution. That is, in this case, NOBUZZ is a poor indicator
of larval density in the rot. Further in both sets of data,
no account is taken of the size of the rot, so that larval
numbers per unit of rot may be quite different for rots producing similar numbers of *D. buzzatii* adults. The cladode
rots in *O. stricta* are often small and short-lived, so that
density-dependent selection would appear possible and worth
further investigation.

   The Collection 45 and coastal and inland rots analyses
both showed significant associations of genetic variables
(for *Est-1*, *Est-2*, *Hex* and *Adh-1*) with yeast species isolated
from the rot. These associations presumably are mediated
through metabolic products of the yeasts. Laboratory and
field experiments have shown that larval and adult *D. buzzatii*
can discriminate among yeast species, and the field experiment indicated that *Est-2* genotypes are differentially
attracted to the yeast species (Barker *et al.*, 1981a, b).

Variation due to metabolic products of the yeasts may well contribute to fitness differences among genotypes and the maintenance of polymorphism.

The strongest associations detected were those with the alcohol concentrations, with very high loadings for methyl alcohol and ethyl alcohol relative to all other variables. Such associations with *Adh-1* would not have been surprising, but all loci except *Pgm* appear to be involved. For loci other than *Adh-1* and *Aldox*, these associations possibly reflect effects of other organic compounds whose concentrations are correlated with those of the alcohols. However, the alcohols and any other organic compounds must be products of the rot microflora (including bacteria as well as yeasts), and dependent on the chemical composition of the cactus. Thus it seems that we are just at the beginning of evaluating the interactions among the components of the cactus-yeast-*Drosophila* system in relation to understanding the population genetics of allozyme polymorphism in the cactophilic *Drosophila*.

ACKNOWLEDGMENTS

I am most indebted to Peter East for his contributions to all aspects of these studies, to Professor H.J. Phaff and Ms M. Miranda for identification of yeast species, to Professor M.G. Pitman, School of Biological Sciences, University of Sydney for provision of facilities and to Marcia Corderoy for assistance with the analyses of ion concentrations, to Professor G. Starmer, Department of Pharmacology, University of Sydney for the alcohol analyses, and to Merrilee Baglin and Bonita Moss for technical assistance.

REFERENCES

Barker, J.S.F. (1977). *In* "Measuring Selection in Natural Populations" (F.B. Christiansen and T. Fenchel, eds.), Lecture Notes in Biomathematics *19*, 403. Springer-Verlag, Berlin.

Barker, J.S.F. (1981). *In* "Genetic Studies of *Drosophila* Populations" (J.B. Gibson and J.G. Oakeshott, eds.), p.161. Proceedings of the 1979 Kioloa Conference, The Australian National University, Canberra.

Barker, J.S.F., and East, P.D. (1980). *Nature 284*, 166.
Barker, J.S.F., and Mulley, J.C. (1976). *Evolution 30*, 213.
Barker, J.S.F., Parker, G.J., Toll, G.L., and Widders, P.R. (1981a). *Aust. J. Biol. Sci. 34*, 593.
Barker, J.S.F., Toll, G.L., East, P.D., and Widders, P.R. (1981b). *Aust. J. Biol. Sci. 34*, 613.
Christiansen, F.B., Frydenberg, O., Hjorth, J.P., and Simonsen, V. (1976). *Hereditas 83*, 245.
East, P.D. (1982). Chapter 21, this Volume.
Hedrick, P.W., Ginevan, M.E., and Ewing, E.P. (1976). *Annu. Rev. Ecol. & Syst. 7*, 1.
Heed, W.B. (1978). *In* "Ecological Genetics: The Interface" (P.F. Brussard, ed.), p.109. Springer-Verlag, New York.
Jones, J.S. (1980). *Nature 286*, 757.
Karlin, S. (1981). *Genetics 97*, 457.
Lewontin, R.C., Ginzburg, L.R., and Tuljapurkar, S.D. (1978). *Genetics 88*, 149.
Mather, W.B. (1957). *Univ. Texas Publ. 5721*, 221.
Mulley, J.C., and Barker, J.S.F. (1977). *Drosophila Inf. Serv. 52*, 151.
Mulley, J.C., James, J.W., and Barker, J.S.F. (1979). *Biochem. Genet. 17*, 105.
Murray, N.D. (1982). Chapter 2, this Volume.
Oakeshott, J.G., Chambers, G.K., East, P.D., Gibson, J.B., and Barker, J.S.F. (1982). *Aust. J. Biol. Sci. 35*, 73.
Powell, J.R., and Taylor, C.E. (1979). *Amer. Sci. 67*, 590.
Vacek, D.C. (1982). Chapter 12, this Volume.
Wagner, R.P. (1944). *Univ. Texas Publ. 4445*, 104.
Wasserman, M. (1982). Chapter 4, this Volume.
Zouros, E. (1973). *Evolution 27*, 601.

15
Life History Evolution Under Pleiotropy and K-Selection in a Natural Population of *Drosophila mercatorum*[1]

Alan R. Templeton

Department of Biology
Washington University
St. Louis, Missouri

J. Spencer Johnston

Department of Plant Sciences
Texas A&M University
College Station, Texas

I Introduction

Wright (1977) has forcefully argued for the assumption of "universal pleiotropy" with respect to genetic variants influencing fitness. Fitness, of course, is a function of the life history properties of the individual; that is, age-specific viabilities and fecundities. The assumption of universal pleiotropy with respect to life histories therefore means that the alteration of one life history parameter simultaneously induces changes in one or more other such parameters. These pleiotropic constraints are critical to a proper understanding of life history evolution, but have traditionally been very difficult to explore analytically unless the dimensionality of the problem is severely limited. However, Templeton (1980, 1982a) has recently developed a

[1]*This research was supported by NIH grant 5RO1 AGO2246-03.*

ECOLOGICAL GENETICS AND EVOLUTION
ISBN 0 12 078820 9

generalized life-history theory under density-independent and density-dependent growth that can deal with an arbitrary number of pleiotropic modifications in an analytically tractable fashion. One very important conclusion emerging from this theory was the critical role played by pleiotropy *per se* in determining life-history evolution.

In this paper, we first describe a genetic system with extensive pleiotropic effects that is polymorphic in a natural population of *Drosophila mercatorum*. Next, we describe the ecological consequences of a severe drought that occurred in 1981 and, finally, we relate the observed genetic alterations accompanying this drought to the life history theory developed by Templeton (1982a).

II   Abnormal Abdomen

Natural populations of *Drosophila mercatorum* from the Island of Hawaii are polymorphic for a trait known as abnormal abdomen (*aa*) (Templeton and Rankin, 1978). The primary cause of *aa* appears to be a 6 kilobase insertion into the 28s ribosomal gene (DeSalle, personal communication). X-chromosomes can be classified as either allowing the expression of *aa* or not (Templeton and Rankin, 1978); thus, for the purposes of this study *aa* can be regarded as a single, X-linked "locus" with two alleles (*aa* and +). Normally the expression of *aa* is limited to females, but certain Y-chromosomes allow the expression of *aa* in males provided the males also have an *aa* X (Templeton, unpublished data). These *aa* Y's have apparently undergone a deletion of some ribosomal DNA normally found in the Y-chromosome (DeSalle, personal communication). There are also several autosomal modifiers of *aa* (Templeton and Rankin, 1978), but we will not deal with them here other than to state that they are also polymorphic in the natural population.

Given the fundamental importance of ribosomes for protein synthesis, it is not surprising that *aa* has many pleiotropic effects. The syndrome is named for the fact that juvenilized cuticle is retained in the adult, causing a disruption of normal segmentation, pigmentation and bristle formation. This attribute, along with a prolonged larval development, is temperature sensitive, being most extreme at 28^0C, less extreme at 25^0C, and totally absent at 18^0C. Since the natural

population inhabits an area where daytime temperatures are
normally between 15^0-22^0 and the nights are cooler, it is
doubtful if these pleiotropic consequences are important in
nature. However, *aa* also has an impact on adults. It
decreases adult longevity, but increases the rate of ovarian
maturation and egg production, even in heterozygotes
(Templeton, 1982b & c). Consequently, the *aa* and + alleles
represent a classic life history trade-off; fecundity vs.
viability.

As mentioned earlier, *aa* is polymorphic in a natural
population of *Drosophila mercatorum* found on the island of
Hawaii near the town of Kamuela. *D. mercatorum* is a *repleta*
group *Drosophila*, and like most members of this group is
cactophilic. The distribution of *D. mercatorum* coincides
with the distribution of the cactus *Opuntia megacantha* which
was introduced around 1809 from Acapulco, Mexico. It is
likely that *D. mercatorum* was introduced to Hawaii with this
cactus, as was *D. hydei*, another *repleta* group *Drosophila*
found in these cacti. Both species can be collected in large
numbers in the cactus patches, but are totally absent in
collections made in nearby areas where the cacti are absent.
Also, both species can be reared out of rotting pads of the
cactus.

Individual cacti can become quite large in *Opuntia
megacantha*. For example, cactus B-1 in one of our study
areas encompasses a volume of 367 cubic meters. In addition,
very humid air (relative humidities are generally from the
upper 70% to about 95%) is continually blown through our
prime study sites most of the year. As a consequence, each
individual cactus normally has a large number of rotting pads
either on it or on the ground underneath, and rather large
populations of flies can be supported by an individual plant.
For example, the B-1 cactus in May, 1980 supported a
population of 971 ± 76 (standard deviation) *D. mercatorum*
and 708 ± 86 *D. hydei* as estimated by a three-capture,
log-linear model; Heckel and Roughgarden, 1979; marking
techniques are described in Johnston and Templeton (1982).
The nearby cactus B-2 (enclosing a volume of approximately
215 cubic meters) had 626 ± 126 *D. mercatorum* and 494 ± 76
D. hydei. These estimates are possible because the flies do
not disperse even between adjacent cacti while the trade
winds are blowing but confine dispersal to days of little or
no wind (i.e., wind speeds of less than 11 kilometers per
hour). Because the trade winds blow most of the year (for
example, in 1979 at the nearby Kamuela airport the trades
blew all but 49 days, and 31 of these still days occurred in

the months November through April), each cactus represents
an isolated subpopulation that often exists for several
generations with no gene flow in or out.

The cacti are distributed on the leeward side of the
Kohola mountains, ranging from an elevation of about 800
meters to 1100 meters at our prime study site located near
Puu Kawaiwai, an extinct cinder cone. We have established
five principal collection areas (lettered A through E) along
this gradient, with A being at the top and E at the bottom.
In addition, we have also studied a second side (site IV)
about 3 km away from Puu Kawaiwai and at an elevation of 670
meters above sea level. This site is warmer and less humid
than the Puu Kawaiwai sites.

The allele frequency of *aa* in the B-1 cactus population
has been monitored since 1976 by crossing wild caught males
to a laboratory strain homozygous for *aa* and with a
background that maximizes penetrance of the abdominal
abnormalities, and/or by crossing the male offspring
obtained from wild-caught females (the females had mated in
nature) with this same homozygous *aa* strain. If these males
bore an *aa*-X, their daughters in this cross will display
abdominal abnormalities and, independently of whether or not
the males have an *aa*-X, their sons in these crosses will
express *aa* if they have an *aa*-Y. In this way, we can assay
for both the frequencies of *aa* X's and *aa* Y's. Finally,
before pooling the frequency of *aa*-X's from wild-caught males
with those scored from paired sons of wild-caught females, it
is necessary to regard a sample of 2N sons (N pairs) as a
sample of 3N/2 independent X-chromosomes. Likewise, we
correct for non-independence of the Y-chromosomes within a
tested pair by regarding each pair of sons as yielding
information about only a single Y-chromosome unless the pair
is discordant.

The allele frequencies of *aa*-X in the B-1 population were
.225 (N = 40) in 1976, .220 in 1978 (N = 100) and .275 in
1980 (N = 39). None of these differences are significant and
yield a pooled frequency of 0.233 (N = 179). Thus, from 1976
through 1980, there seems to be an equilibrium situation with
respect to *aa*-X's (unfortunately, *aa*-Y's were not scored
during this time period). However, there is strong evidence
for spatial variability in *aa*-X frequencies. In May, 1980,
the frequency of *aa*-X at site IV was .455 (N = 11), and in a
collection made in January, 1981 it was .382 (N = 34). These

differences are not significant and yield a pooled estimate
of 0.400 (N = 45). Hence, the frequency of *aa*-X is nearly
twice as high at the drier, lower elevation site, and this
difference is significant at the 5% level. In addition, the
frequency of *aa*-Y was determined for site IV to be .121 (N =
33) during January, 1981. This sample at site IV was taken
just at the start of the worst drought experienced in this
area since weather data have been collected. We also
obtained genetic and ecological data in July and August,
1981, during the drought. We will now discuss the ecological
and genetic impact of this drought.

III   The Ecological Impact of the 1981 Drought

The most obvious and immediate impact of the drought was
a severe reduction in population size. This was evident from
just the difficulty of collecting a handful of specimens
during 1981 in contrast to the ease of collecting more
individuals than we could handle during 1980. However, a
more quantitative measure of this decline in population size
is possible.

Using a single marked-recapture, we directly estimated
the population sizes found in four cacti during the drought.
The estimates of population size are shown in Table I. As
can be seen from this table, the B-1 cactus population has
had an 81% reduction in population size relative to its
pre-drought size. However, our collection data indicated
that other cactus populations suffered even more severe
reductions in population size. To quantify this field
impression, we first assume that the population sizes in the
various cacti we were collecting are proportional to our
total catches in that cactus. We can test the adequacy of
this hypothesis by assuming it is true for the four cacti in
which we directly estimated population size. The bottom row
of Table I gives the estimated population sizes for these
four cacti that retain the actual collection proportions and
were adjusted to minimize the squared deviations from the
hypergeometric estimates. As can be seen, all the resulting
estimates are approximately within one standard deviation
unit of the hypergeometric estimates. Hence, this assumption
seems to be a good one. Moreover, the bottom row of Table I
gives us an absolute number by which to adjust the population

TABLE I.   Estimated Population Sizes of Drosophila
mercatorum in Four Cactus Patches During the
Drought of 1981.   The Estimates are Based upon
a Marked-Release, Recapture Experiment Assuming
a Hypergeometric Sampling Distribution Cor-
rected for Dispersal.   The Hypergeometric Esti-
mates are Given (N), as well as the Standard
Deviations (s.d.) of Those Estimates.   The
Bottom Row Tests the Assumption that Population
Sizes are in the Same Proportion as Total
Captures.   The Estimated Sizes in this Row (N)
are in Exactly the Same Proportion as the Total
Captures in These Four Cacti, with the Absolute
Values of N's being Fixed by Finding the Mini-
mum Mean-Squared Deviations from the N's that
Preserve the Total Capture Proportions

Cactus	B-1	C-1	D-1	D-5
$N \pm$ s.d.	187 ± 50	179 ± 44	56 ± 31	62 ± 25
N	240	192	31	63

size estimates in all the other cacti under the assumption
that size is proportional to actual captures. The resulting
estimates for 31 cactus populations surveyed in our dispersal
study (Johnston and Templeton, 1982) are given in Table II.
Note that the estimated size of cactus B-2 population is
only 24 *D. mercatorum*. This represents a 96% reduction in
population size relative to the pre-drought levels. The
reason for this difference probably lies in microgeographical
variables. B-1 lies just downwind from and actually abuts
on a road embankment. Moreover, the cactus itself is
extremely large and has a dense, almost wall-like, unbroken
upwind exterior (the Hawaiian name for these cacti is
panini--an "unfriendly wall"). This shields the interior
of the cactus (where the flies live) from the dessicating
winds. On the other hand, B-2 is totally exposed to the wind
and has a much less dense exterior that allows more of the
wind to blow through it. Under pre-drought conditions, tall
grass helps protect B-2 from the wind, but this protection
was lost during the drought. This seems to be a general
pattern. Therefore, the drought not only reduced the total

TABLE II.   Estimated Population Sizes of Drosophila
mercatorum for 31 Cactus Patches During the
Drought of 1981. The Cacti are Arranged by
Area (A through D) and by Elevation, Going
from Higher to Lower Elevations

Cactus	$\hat{N}$	Cactus	$\hat{N}$	Cactus	$\hat{N}$
A-2	2	C-3	41	C-29	29
A-1	9	C-7	23	C-30	12
		C-6	54		
B-10	0	C-33	14	D-7	8
B-4	6	C-4	6	D-4	28
B-9	0	C-1	192	D-6	14
B-1	240	C-8	29	D-5	63
B-2	24	C-24	26	D-1	31
		C-34	24	D-8	29
C-31	55	C-26	8	D-14	27
C-32	5	C-12	15	D-15	26
		C-35	15		

population size, but did so in a highly non-homogeneous
fashion such that the bulk of the population was now being
supported by a handful of cacti. The drought therefore
accentuates the population subdivision by making the
effective distances between individuals much larger.

The drought also had a severe effect on individual adult
viability. The trade winds began blowing on the afternoon
of our first day recapture in our dispersal experiment, and
there was no further dispersal during the remainder of the
study (one week). Consequently, from day one on, the
proportion of our sample that was marked could only decrease
due to mortality, and if the population size were constant
over this interval, the rate of decrease in the proportion
of marked individuals would be a direct estimate of the
mortality rate.

We regressed these proportions against time and obtained
a significant regression for *D. mercatorum* but not *D. hydei*.
The regression indicates that 19% of the *mercatorum* adults
died per day during the drought. This estimate of mortality
will be an overestimate if the total population size were
growing during this period and an underestimate if it were

declining. Since the drought was in full force during our dispersal study, we deem it unlikely that the population size was growing during this study, so if anything, 19% is an underestimate of daily adult mortality rates of *D. mercatorum*. During 1980, we had not done any long term dispersal experiments, but we did estimate age structure (Johnston and Templeton, 1982). The age structure in 1980 had many old flies at the upper elevations, thereby indicating very low daily mortality rates under pre-drought conditions. Thus, we conclude that the drought greatly increased adult mortality in *D. mercatorum*.

Another effect of the drought was to reduce the number of available oviposition sites for adult females. As mentioned earlier, rotting pads are the larval food substrate. In 1980, such rots were very easy to find and contained many flies. In 1980, we bagged a total of 14 rots to monitor emergence patterns. Over the next 17 days, ten (71%) pads produced emerging *D. mercatorum,* and six of these (43% of total pads) yielded *D. hydei.* In these six, *D. mercatorum* tended to emerge earlier than the *hydei.* Of the pads yielding *D. mercatorum,* the numbers ranged from 1 to 31 with an average of 9.1. The comparable figures for *D. hydei* were a range of 1 to 22 with an average of 8.0. The emergence window was 14 days for *mercatorum* (i.e., maximal number of days separating first from last emergence for a given rot) and 17 days for *hydei.* During the drought, we bagged 18 rots, using the same criterion we did in 1980 for choosing pads to be bagged. Over the next 19 days, only two (11%) rots produced flies, both yielding *mercatorum* and only one yielding *hydei,* with an average of 3.5 *D. mercatorum* and seven *D. hydei.* Consequently, for *D. mercatorum* the number of available rots suitable for larval development was greatly reduced from pre-drought levels, and the number of adults emerging from a rot was also reduced. In addition, the window of emergence was reduced from two weeks to three days. The reason for this is straightforward; under drought conditions rots simply dry out much faster and therefore are available as a suitable larval substrate much less of the time.

We also have evidence that the larvae in these rots were under severe stress. *D. mercatorum* is normally about the same size as *D. hydei,* and both are much larger than *D. melanogaster* or *D. simulans.* However, in our collections during the drought, we often caught small flies which we initially thought were *D. simulans* based on size alone, but

under closer inspection proved to be *D. mercatorum*. 5.5% of
the *D. mercatorum* males collected (out of a sample of 91)
and 7.0% of the females (N = 57) were miniature. Such
miniature adults had never before been observed in the
natural population, but have been observed emerging from
old laboratory cultures where the food was almost completely
exhausted or mouldy. Thus, the high occurrence of miniature
adults strongly indicates larval stress during the drought
that was absent in pre-drought conditions.

IV The Genetic Impact of the 1981 Drought

The drought also had dramatic genetic consequences upon
the frequencies of aa-X and Y chromosomes. Table III gives

TABLE III.   *Estimated Frequencies of aa-X's and aa-Y's in
Populations of Drosophila mercatorum During
the 1981 Drought. Estimates from Wild-Caught
Males and Paired Sons of Wild-Caught Females
are Pooled in this Table. Each Pair of Sons
is Regarded as Giving Information on 3/2
X-Chromosomes and 1 Y-Chromosome (unless
Discordant). Hence, Sample Sizes for aa-X
Frequencies can Take on Integer and/or Half
Values*

Location	Frequency aa-X	Sample size	Frequency aa-Y	Sample size
A-1	0.000	2.5	0.500	2
B-1	0.486	37	0.121	33
C-1	0.353	29	0.217	23
C-6	0.279	47.5	0.240	25
D-5	0.500	22	0.059	17
D-15	0.675	10	0.000	7
E	0.406	26.5	0.130	23
IV	0.438	12	0.143	7

the frequencies observed at cacti from sites A through E and at site IV. Concentrating initially upon the Puu Kawaiwai sites A through E, we can see there is considerable heterogeneity between cactus populations in the frequency of *aa*-X's and *aa*-Y's. To see if this heterogeneity is real or just an artifact of our sample sizes, we transformed the observed allele frequencies with

$$a_i = \arcsin \sqrt{x_i/n_i} \qquad (1)$$

where x_i is the number of *aa*-X's (or Y's) in a sample of n_i chromosomes from cactus i. The advantage of this transformation is that the sampling variance of a_i equals $1/(4n_i)$ and therefore does not depend upon the allele frequency. Now consider the statistic

$$4 \sum_{i=1}^{r} n_i (a_i - \bar{a})^2 \qquad (2)$$

where $a = \Sigma n_i a_i/n$, $n = \Sigma n_i$, and r is the number of populations sampled. Under the null hypothesis that the allele frequencies are identical in all cactus populations and that all differences in observable a_i's are due to sampling error alone, the expectation of statistic (2) is (r-1) and from large sample size theory, the asymptotic distribution of (2) is chi-square with r-1 degrees of freedom. This statistic therefore tests whether or not there is significant variation between cactus populations with resepct to allele frequencies. Applying this statistic to the data on sites A-E given in Table III yields a chi-square of 13.21 with six degrees of freedom for the *aa*-X's (significant at the 5% level) and 9.43 with six degrees of freedom for the *aa*-Y's (not significant at the 5% level). Therefore, there is a significant heterogeneity for *aa*-X's. A possible explanation for an absence of the heterogeneity for *aa*-Y's will be given later. The next question of interest is whether or not this heterogeneity is due to selection, drift, or a combination of both. This is particularly important in light of our dispersal data. The flies sampled for allele frequencies were collected on August 1 through August 4, 1981 (Johnston and Templeton, 1982). The trade winds had died down to virtually no wind on July 19, and 31% of the *D. mercatorum* dispersed to other cacti. Indeed, we directly observed exchange of individuals between several of the cacti listed in Table III. Moreover, the wind also dropped to near zero on July 29 and again on

July 31. Although we were not monitoring dispersal directly
at sites A-E on these later two dates, we would expect that
dispersal occurred then as well. Consequently, the
populations listed in Table III are not isolated from one
another. However, as the figures in Table II indicate, the
actual population sizes were quite small. Therefore,
consider the null hypothesis that the population is truly
panmictic and that the heterogeneity is induced solely by the
small numbers of organisms inhabiting any particular cactus.
Let N_i = the actual population size at cactus i. The
expectation of statistic (2) under this null hypothesis is

$$[(r-1) + n_i/N_i] \tag{3}$$

Thus, given the N_i's, the statistic

$$4\Sigma n_i (a_i - \bar{a})^2 - \Sigma n_i/N_i \tag{4}$$

is asymptotically distributed as a chi-square with r-1
degrees of freedom. It is important to note that this
statistic tests for the heterogeneity induced by subdivision
into finite sized subpopulations in an otherwise panmictic
population and *not* for the expected sampling heterogeneity
induced by sampling error. Our estimates for the N_i's for
all cacti but E are given in Table II. Once again, based
upon our collection experiences at E compared to the earlier
collections up at A-D, E has approximately 100 flies. However, before using these figures we note that a population
of N_i individuals corresponds to a population of $3N_i/2$
X-chromosomes under the assumption of a 50:50 sex ratio
(which holds for the *D. mercatorum* population). Thus, the
second sum of statistic (4) becomes $\Sigma 2n_i/(3N_i) = 1.64$ and
statistic (4) takes on the value of 11.57. This is not
significant at the 5% level, indicating that the population
distributed over A-E is panmictic. A similar test can be
performed with the Y-chromosome data, except there the
second sum is $\Sigma 2n_i/N_i$. The Y-chromosome statistic (4) has a
value of 6.01.

   The above analysis implies that the populations inhabiting sites A-E are panmictic with respect to *aa* and that the
differences in allele frequencies observed between cacti can
be explained solely due to small population sizes and do not
require any selective differences. However, the data in
Table III also give evidence of very strong selection
operating upon *aa*. First, continue to confine attention to
sites A-E. The ranking of cacti with respect to *aa*-X

frequency is D-15 > D-5 > B-1 > E > C-1 > C-6 > A-1.   The
ranking of cacti with respect to *aa*-Y frequency is exactly
the opposite; that is, there is perfect rank correlation of
-1 between the frequencies of *aa*-X's and *aa*-Y's.   This result
is significant at the 1% level.   How do we make sense of
this result in light of the evidence discussed earlier for
extensive dispersal of both males and females in this area?
Recall that the *aa*-X's and *aa*-Y's are determined by
examination of the X and Y chromosomes borne by wild-caught
males and by examination of the X and Y chromosomes borne by
the sons of wild-caught females.   Consequently, all the X
chromosomes scored represent X's that were present in
individuals caught in a particular cactus at the time of the
sample, but only the Y's scored from wild-caught males
represent Y chromosomes that were present in individuals
caught in the cactus.   The Y chromosomes scored from the sons
of wild-caught females represent Y chromosomes found in the
sperm loads of the wild-caught females, and therefore are
indicative *not* of what type of males are in that cactus at
the time of sampling, but of what males the females had
previously mated with.   This effect probably also explains
the lack of significant heterogeneity for the Y's.
Separating our data into Y's scored from wild-caught males
and Y's scored from sons of wild-caught females, the Spearman
rank correlation of the frequency of *aa*-X's with *aa*-Y's in
wild caught males is -0.21, which is not significantly
different from zero.   This indicates that there is no
association between the spatial distribution of individuals
bearing *aa*-X's with those bearing *aa*-Y's.   On the other hand,
the Spearman rank correlation of frequency of *aa*-X's with
aa-Y's found in the sperm load of wild-caught females was
-0.94, which is significant at the 1% level.   This indicates
very strong disassortative mating with respect to abnormal
abdomen.   Thus, we have evidence for very strong sexual
selection upon this phenotype.

  Table III also contains evidence for strong natural
selection.   Note that the frequency of *aa*-X's and *aa*-Y's at
site IV are .437 and .14 respectively.   These are not
significantly different from the pre-drought estimates of
.400 and .121 and yield pooled estimates of .408 (N = 57)
for *aa*-X and .125 (N = 40) for *aa*-Y's.   However, at cactus
B-1 the pre-drought frequency of *aa*-X's was .233 from
1976 through 1980 which is significantly different from the
drought frequency of .487 at the 0.1% level.   Moreover, since
we have evidence that sites A-E represent a single panmictic
and homogeneous population with respect to *aa* frequency, the

pooled A-E frequency of *aa*-X's is .401 (N = 174.5), which is
also highly significantly different from the pre-drought B-1
frequencies. However, neither the B-1 *aa*-X frequency nor
the A-E pooled frequency is significantly different from any
of the site IV frequencies or the pooled site IV frequency.
Likewise, there are no significant differences between the
frequency of *aa*-Y's at site B-1 or the pooled A-E sites from
site IV. Therefore, what we have is a dramatic two-fold
increase in the frequency of *aa* during the drought from the
pre-drought levels in the B-1 population, but no change at
all in the site IV population. Recall that site IV is in a
location that is normally much drier and less humid than
sites A-E. Thus, the drought has caused the allele
frequencies of *aa* to converge to the dry site values
throughout the entire population. This fact coupled with
the previous stability of *aa* frequencies at cactus B-1 over
several years is very strong evidence for natural selection.

V   Abnormal Abdomen, Pleiotropy, and K-Selection

Can the dramatic increase in the frequency of *aa* be
explained in light of our current knowledge of the
pleiotropic consequences of *aa* and of the ecological impact
of the drought? We will now attempt to do so in terms of
the model given in Templeton (1982a). If the pre-drought
population were in equilibrium, the theory of Templeton
(1982a) indicates that

$$\sum_{i=a}^{n} \ell_{i+1}w_{g,i+1} = \sum_{i=a}^{n} v_{g,i} \sum_{j=i+1}^{n+1} \ell_j m_j \tag{5}$$

when the ℓ_i's are the probabilities of surviving to age i,
the m_i's are the age-specific fecundities, the v's are the
viability deviations induced by genotype g at age i, and the
w_{gi}'s are the fecundity deviations induced by genotype g at
age i. Since *aa* decreases longevity but increases
fecundity, the v's will be negative and the w's positive for
aa. As we go into the drought conditions, this balance will
be altered, and the fate of *aa* will be determined by which
of the following two quantities is larger:

$$\sum_{i=a}^{n} \ell'_{i+1} f_{i+1}(eq) w_{g,i+1} \tag{6}$$

vs.

$$- \sum_{i=a}^{n} v_{g,i} \sum_{j=i+1}^{n+1} \ell'_j m_j f_j(eq) = - \sum_{i=a}^{n} \ell'_i v'_i v_{g,i} \tag{7}$$

where ℓ'_i is the probability of surviving to age i under drought conditions, the f's represent the decrease in oviposition sites induced by the drought, and v' is the reproductive value of age i under drought conditions. If quantity (6) is larger than (7), we would expect the frequency of *aa* to increase in the population, and if the reverse were true it should decrease. Note that both the fecundity effects of *aa* (the w's) and the viability effects (v's) are weighted by the ℓ' terms, which correspond to the stable age distribution under drought conditions. Given our exceedingly large daily mortality under drought conditions, this means that only the first few terms in either (6) or (7) will have any major impact on fitness under drought conditions. Thus, the earlier ovarian maturation of *aa* will favor (6) over (7). In addition, the w's are also weighted by the f's, and hence the drought induced decreases in effective fecundity are going to reduce this contribution to fitness. However, the v's are weighted by the reproductive values; that is, the expected number of *future* offspring. This weighting factor is severely reduced both by the extremely high rates of daily mortality and by the restricted number of oviposition sites. Reproductive values are therefore going to be quite small for all ages. Hence, under very broad conditions, quantity (6) will exceed quantity (7); that is, dry conditions should selectively favor *aa*, which they apparently did. This prediction also explains the pre-drought cline in *aa* frequency.

Note that the drought conditions severely reduced the carrying capacity of the environment. Under traditional concepts of r- and K-selection, a severe reduction in carrying capacity implies a tremendous increase in "K-selection". Yet, this severe imposition of "K-selection" favored, both in theory and in reality, the *aa* allele which causes phenotypes displaying classic "r-selected" traits--an increased in fecundity, faster ovarian maturation, and

decreased longevity. These facts illustrate that when
pleiotropic constraints are introduced into life history
evolution, the standard "optimal" life histories and the
straightforward dichotomies between "r-selected" and
"K-selected" traits all vanish. Thus, an understanding of
the genetic basis of the traits under selection is
absolutely critical for predicting the types of life
histories that will evolve under certain ecological
conditions. The classic r- and K-framework is simply
inadequate for this task.

REFERENCES

Heckel, D.G., and Roughgarden, J. (1979). *Ecology 60,* 966.
Johnston, J.S., and Templeton, A.R. (1982). Chapter 16, this
    Volume.
Templeton, A.R. (1980). *Theor. populat. Biol. 18,* 279.
Templeton, A.R. (1982a). In preparation.
Templeton, A.R. (1982b). The Prophecies of Parthenogenesis.
    In "Variation in Life Histories: Genetics and
    Evolutionary Processes" (H. Dingle and J. Hegmann, eds.).
    Springer-Verlag, New York. (in press).
Templeton, A.R. (1982c). Natural and Experimental
    Parthenogenesis. *In* "The Genetics and Biology of
    Drosophila" Vol. 3, (M. Ashburner, H.L. Carson and
    J.N. Thompson, Jr., eds.), Academic Press, New York.
    (in press).
Templeton, A.R., and Rankin, M.A. (1978). *In* "The Screw-worm
    Problem" (R.H. Richardson, ed.), p. 83. University of
    Texas Press, Austin.
Wright, S. (1977). "Evolution and the Genetics of
    Populations," Vol. 3. University of Chicago Press,
    Chicago.

16
Dispersal and Clines in *Opuntia* Breeding
Drosophila mercatorum and *D. Hydei* at Kamuela, Hawaii[1]

J. Spencer Johnston

Department of Plant Sciences
Texas A&M University
College Station, Texas

Alan R. Templeton

Department of Biology
Washington University
St. Louis, Missouri

I Introduction

Patterns of geographic distribution of gene frequencies
in populations provide primary information on the population
genetics and natural history of species. It is not suprising therefore that patterns of gene frequency and in particular of gradients in gene frequency (clines) have been
the subjects of much recent attention. Naturally occurring
gene frequency clines have been widely reported for chromosomal types and electromorphs, and the theory of clines has
been the subject of recent study (Barton, 1979; Nagylaki
and Lucier, 1980; Slatkin, 1981). Correlations of clinal
variation with weather and habitat variables have been reported, yet as pointed out by Endler (1977), it is impossible to interpret a natural cline without knowing the shape
of the fitness curve across the habitat and the extent of

[1]*This work was supported by NIH Grant 5R01 AG02246-03.*

ECOLOGICAL GENETICS AND EVOLUTION
ISBN 0 12 078820 9

gene flow. Further, with a few important exceptions (for
example, Parsons and McKenzie, 1972), it has been impossible
to discriminate between clines caused by selection in an area
of primary contact and those resulting from reduced dispersal (with or without selection) in an area of secondary
contact.

In *Drosophila* populations, geographical clines generally
involve areas so large that the effects of ecological factors
and history are hard to tease apart. This problem is made
greater by the very large dispersal of *Drosophila* during
unfavorable conditions (Jones *et al.*, 1981) and by the
observation that drift among neutral alleles is obviated by
exchange of even one individual per generation per population
(Felsenstein, 1976). In the few studies where microgeographic distributional differences have been observed and
where the dispersal and habitat parameters have been determined (Richardson and Johnston, 1975; Taylor and Powell,
1978), reduced dispersal associated with habitat selection
has been of critical importance to the maintenance of the
cline. In these very promising studies, a description of the
ecological factors permitting habitat selection and a measure
of the fitness sets within each microhabitat provide valuable
information on all aspects of evolutionary and ecological
genetics.

In this regard, a geographic cline reported by Templeton
and Johnston (1982) is of particular interest. Two
Drosophila species, *D. mercatorum* and *D. hydei* overlap
throughout a very restricted range on the leeward slide of
the Kohala Mountains. The introduction of their single,
shared host plant (*Opuntia megacantha*) is known to have occurred in the year 1809, and now *D. mercatorum* exhibits a
cline in frequency of "abnormal abdomen," a syndrome associated with an insertion in the 28s ribosomal gene. First observed in 1976 on a hillside site extending from 670 meters
to 1100 meters near an extinct cinder cone (Puu Kawaiwai),
the cline persisted with little change until 1981, when the
cline disappeared coincident with a drought and a reduction
of population size.

Here, we report on the dispersal rate of *D. mercatorum*
and *D. hydei* and investigate the forces which shape the
cline. Further, we report a cline in the relative abundance
of the two species and show that the same forces influence
both the genetic cline and the species cline.

TABLE I.   *Abundance of D. mercatorum and D. hydei*

Site	May 1980 Totals % merc.		July 1981 Totals % merc.	
A	141 merc.	21%	15 merc.	12%
	520 hydei		113 hydei	
B	361 merc.	68%	485 merc.	65%
	172 hydei		256 hydei	
C	390 merc.	81%	675 merc.	76%
	92 hydei		209 hydei	
E (Site IV)	80 merc.	100%	289 merc.	99%
	0 hydei		4 hydei	

II The Species Cline

Table I shows the 2436 *D. mercatorum* and 1366 *D. hydei*
captured at guava baits and naturally rotting *Opuntia*
cladodes at 70 cacti in four contiguous areas of the Puu
Kawaiwai study site (Figure 1). The total numbers represent
different levels of trapping effort in each area and cannot
be directly compared, but the relative proportion of

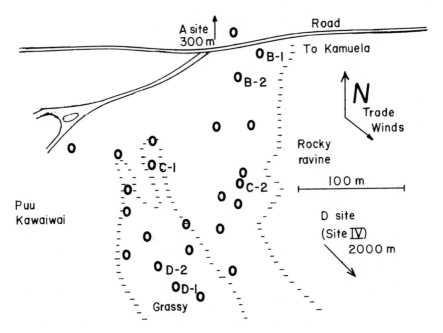

FIGURE 1.   *Puu Kawaiwai collecting sites.*

D. mercatorum, which is independent of effort, varied in 1980
from 21% at the upper site to 100% at the base of the hill.
In 1981, this same cline went from 12% to 99%. As will be
seen, the area of the cline is less than the dispersal
capacity of the species.

A. *Rearing Records*

Flies were reared from 14 necrotic pads from the B site
(Figure 1) in 1980 and from 18 necrotic pads from the B or
E sites in 1981. The pads were collected as they were
encountered beneath the cactus, placed into nylon bags, and
then set into the shade. Daily collection records (Table II)
show considerable heterogeniety among pads.   Total flies
reared per pad ranged between zero and 138. Eleven of 14
pads (79%) were productive in 1980 and 4 of 11 (29%) produced
only *D. mercatorum*. In 1981 a much smaller proportion was
productive (2 of 18 or 11%) and 1 of 2 (50%) produced only
D. mercatorum. All of these emergence rates are very low as
compared with the size of a *megacantha* pad which is 30 to 70
cm across and 4 to 6 cm thick. The total emergence from the
11 productive pads in 1980 was 176 *D. mercatorum* ($\overline{X} = 16 \pm$
7.9) and 143 *D. hydei* ($\overline{X} = 20 \pm 10.6$). The total emergence
for the two pads in 1981 was 2 *D. mercatorum* and 7 *D. hydei*
in one pad and 5 *D. mercatorum* from the other. On the
average *D. hydei* emerged 2.6 days later than *D. mercatorum*
(13.5 ± 1.60 vs. 10.9 ± 1.90, p < 0.01), while total production did not differ between the two species.

III Dispersal

Mark and recapture studies were conducted by aspirating
flies from both naturally occurring necroses and from one
gallon plastic bags containing 3-4 rotted guava, placed
within the larger cacti in the area. On July 18, 1981, flies
captured in sites B-1, B-2, C-1, C-2, D-1, and D-2 (Figure 1)
were marked by shaking them in a vial containing one of six
colors so that flies at each cactus were uniquely marked.
The respective colors used were yellow, red, green, blue,
orange and phospho-green, corresponding to Helecon Fluorescent Pigments, U.S. Radium Corporation, pigment numbers 2267,
2225, 3206, 2205, 2220, and 2330. Marked flies were released
shortly after 12:00 noon, July 18, 1981, when the flies were
least active. On subsequent days recaptures were made at

TABLE II.   *Rearing from Opuntia megacantha Pads*

Date	Pad	*D. mercatorum*			*D. hydei*		
		n	*Mean*	*Variance*	*n*	*Mean*	*Variance*
1980	#1B	10	6.7 *days*	5.5	17	11.2 *days*	5.9
	#2B	58	7.5 *days*	3.1	77	10.9 .*days*	3.3
	#3B	9	3.7 *days*	1.3	36	9.9 *days*	4.3
	#4B	5	10.2 *days*	5.9	0		
	#5B	73	8.8 *days*	2.8	9	14.8 *days*	1.1
	#6B	2	7.5 *days*	5.8	0		
	#7B	2	15.5 *days*	2.1	2	15.0 *days*	2.8
	#8B	1	12.0 *days*	0.0	0		
	#9B	11	15.9 *days*	0.9	0		
	#10B	3	15.0 *days*	0.8	1	16.0 *days*	0.0
	#11B	2	17.0 *days*	0.0	1	17.0 *days*	0.0
	#12-20	None					
1980 Avg.		16.0	10.9[a]*days*	2.7[a]	20.4[a]	13.5[a]*days*	2.5[a]
1981	#1B	2	5.5 *days*	0.7	7	2.4 *days*	1.5
	#2-10	None					
	#1E	5	2.6 *days*	0.6	0		
	#2-8	None					
1981 Avg.		3.5	4.1[a]*days*	0.6[a]	7	2.4[a]*days*	1.5[a]

[a]*Averages do not include pads which produced no flies.*

32 cacti (Figure 1) including the six release sites.   Baits
were opened at 6:30 a.m. and all attracted flies were aspirated off from 7:30 a.m. to 9:30 a.m., after which the bags
were again sealed shut.   Captured flies were taken to the
hotel and sorted without anesthesia by shaking them 10 to 15
at a time into 16 x 8 cm "whirlpac" plastic bags, where they
could be gently held immobile and scored for sex and species
under white light, and then for marking pigments using a
mico-light-I UV-45 ring light (Aristo Grind Lamp Products,
Inc., Port Washington, NY) at 12X magnification on a Wild
M5 dissecting scope.   After scoring all flies were released
at their respective capture sites at 12:00 noon on the
same day.

TABLE IIIa.   Recapture Summary for D. mercatorum

X = number of days after marked flies are released

D_X = proportion dispersed at day X = $\dfrac{\text{\# marked recaptured at day X away from release site}}{\text{total \# marked recaptured at day X}}$

d_X = average distance in meters ($\pm$SD) that flies recaptured away from the release site had moved away

Measure	X=1 7-19	X=2 7-20	X=3 7-21	X=4 7-22	X=5 7-23	X=6 7-24	X=7 7-25	Average
D_X	$\frac{13}{42}=.31$	$\frac{6}{26}=.23$	$\frac{8}{14}=.57$	$\frac{5}{19}=.26$	$\frac{6}{19}=.32$	$\frac{6}{19}=.32$	$\frac{3}{13}=.23$	$\frac{47}{152}=.31$
d_X	67 ± 65	14 ± 7	41 ± 33	55 ± 51	80 ± 82	84 ± 50	84 ± 62	59 ± 56

TABLE IIIb.   Recapture Summary for D. hydei

Measure	X=1 7-19	X=2 7-20	X=3 7-21	X=4 7-22	X=5 7-23	X=6 7-24	X=7 7-25	Average
D_X	$\frac{6}{12}=.5$	$\frac{3}{5}=.6$	$\frac{1}{1}=1.0$	$\frac{6}{6}=1.0$	$\frac{2}{3}=.67$	$\frac{1}{2}=.5$	$\frac{2}{2}=1.0$	$\frac{21}{31}=.68$
d_X	141 ± 53	77 ± 60	544 ± 0	51 ± 26	243 ± 258	244 ± 0	425 ± 168	166 ± 162

The total released *D. mercatorum* at cacti B1, B2, C1, C2, D1, D2 were 97, 12, 116, 34, 24 and 57 respectively, while the total *D. hydei* released at the same site and time were 81, 15, 39, 40, 14 and 92. Total recapture of these same marked populations of *D. mercatorum* were 63, 0, 59, 8, 11 and 10 respectively for a total of 151 recaptures. For *D. hydei* the recaptures were 14, 1, 7, 3, 4 and 2 respectively for a total of 31. The *D. hydei* were recaptured in direct proportion to the release at each cactus. In contrast a χ^2 comparison of *D. mercatorum* releases and recaptures among cacti show heterogeneity ($\chi_5^2 = 22.2$; P < .01) due in large part to the greater proportion of recaptures at the larger more protected B1, C1 and D1 cacti and to a coincidentally smaller proportion of recaptures at the more open exposed B2, C2, D2 cacti. Finally, a comparison of marked but non-dispersant *D. mercatorum* at B1, B2, C1, C2, D1 and D2 (59, 0, 36, 3, 24) against those that moved and were recaptured away from the marking site shows that a significantly greater number of marked flies ($\chi_4^2 = 40.5$; P < .01) remained at the protected B1 and C1 cacti.

Dispersal rates are estimated first as a function of average distance moved and second as root mean square dispersal. The data for *D. mercatorum* are given in Table IIIa. The average proportion moving (D_X) is *47/152* or *.31*, while the average distance moved in meters (d_X) is *59±56*. Sexes are pooled in the table, as no significant differences are seen. The average proportion dispersed among males D_X (♂) is *28/103* or *.27* and among females D_X (♀) is *17/49* or *.35*. The average distance the flies moved is similarly not significantly different between sexes. The average distance moved among males d_X (♂) is *57±53* meters and for females d_X (♀) is *66±63* meters.

Data for *D. hydei* are given in Table IIIb. The average proportion that moved (D_X) is *21/31* or *.68*, while the average distance in meters flies had moved (d_X) is *166±162*. As for *D. mercatorum* above, the data are pooled for sexes because differences are not significant. The average proportion dispersed among males D_X (♂) is *10/17* or *.59* and for females D_X (♀) is *11/14* or *.79*. The average distance moved among males d_X (♂) is *170±100* meters and for females d_X (♀) is *163±142* meters.

A. Weather

The weather information is critical to understanding the
behavior of the flies during this period. Kamuela is
generally windy, cool, and wet, as shown for 1980 and 1981
in Table IV. Temperatures at the time of peak dispersal at
7:30 a.m. were 17-19°C±1°C, and the relative humidity was
generally high, 77-83%. A cline in temperature and humidity
exists across the hill and the top of the hill is generally
in the clouds, cool, and wet, while the lower hillside is
drier and warmer. The temperature and humidity correlate
with the species cline, but dispersal is sufficiently great
that we feel it would be premature to state that temperature
and humidity are important selective agents, until the
critical temperature and humidity dependent life history
parameters are measured. Further, the worst drought in
recorded history occurred in the Kohalas in June and July
1981. The dry hot daytime conditions that resulted are
reflected in a 5°C temperature increase and a 32% humidity
drop for site C at 2:00 p.m. Yet, the species cline did not
change, although the drought raised midday temperatures and
lowered midday humidities by amounts well in excess of the
normal range across the cline.

The dominant weather variable at Puu Kawaiwai is wind.
Trade winds blow continually from the northeast down the
slope at an average velocity of 15 km/hr, and gusts over 35
km/hr are frequently encountered. Exceptional days occur an
average of 4-5 per month. The exceptional days are given a
local name "KONA WEATHER" to indicate that the trade winds

TABLE IV.   *Weather Averages for May-June 1980 and*
July 1981 at Puu Kawaiwai Collection Sites.
The Effect of Drought in 1981 is Evident
from 2:00 p.m. Measurement at Site C

Site	Time	1981			1980		
		Wind	Temp	R.H.	Wind	Temp	R.H.
B	7:30 am	14-20	17±1	83±5%	12-18	18±1	79±9%
C	7:30 am	11-18	18±1	84±6%	0-5	20±1	92±4%
	2:00 pm	11-18	26±1	58±9%	0-5	21±1	90±5%
D	7:30 am	5-14	18.5±1	78±5%			
E	7:30 pm	8-20	19±1	77±5%			
	2:00 pm	12-25	28±1	56±6%			

have stopped and the weather is influenced by the westerly or KONA side of the island. KONA weather existed for the release period and for the first recapture day, July 18 to 19, 1981. The tradewinds picked up and blew 15-20 km/hr down the slope on subsequent recapture days, and greatly limited dispersal after the initial day.

B. *Wind Response in D. mercatorum and D. hydei*

It had earlier been shown (Johnston, 1982) that *D. mercatorum* has a high heritability for wind response measured as crawling with or against the wind passing through interconnected tubes. Further, the wind response in the tubes agreed well with field behavior (Richardson and Johnston, 1975; Johnston, 1978). Here, *D. mercatorum* and *D. hydei* were compared in similar experiments at wind speeds of 0, 2, 4, 6 and 8 km/hr (Table V), at which speeds they crawl actively about the tubes. At speeds of 10 km/hr and higher both species cease moving and hold on to the glass tubes until they are dislodged and blown into the downwind tubes. Eighty wind response trials were run, with 60 or 95% humidity, 2.5 or 5.0 foot candles light levels, and with separate runs for males and females. The two species behave significantly differently, with *D. hydei* moving significantly more into the wind than *D. mercatorum* at all wind speeds between 0 km/hr and 5 km/hr. Neither light, humidity nor sex of fly had a significant effect on direction or extent of movement.

TABLE V.   *Mean Number of Decisions to Move with the Wind (+) or Against the Wind (-) by 30 Flies. Each Value is an Average Over 4 Replicates. Both D. hydei and D. mercatorum were from the Most Windward Collection Site A. Females and Males were Run Separately but are Pooled here as No Differences were Seen*

		Wind Speed in miles per hr					
Collection Site	Species	0	2[a]	4[a]	6	8	Avg.[a]
A	Hydei	.18	-.68	-0.19	0.94	1.33	.31
A	Mercatorum	.22	.44	1.01	1.13	.97	.76

[a]*Denotes species differences at the 5% level.*

In further comparison of 1980 *D. mercatorum* from sites A, B, and C with *D. mercatorum* from site E the flies from the uphill A, B, and C sites were relatively more upwind responsive than those from the site E at the base of the hill (Johnston, unpublished), only because the *D. mercatorum* from E ceased moving at speeds above 6 km/hr and held onto the glass until dislodged and blown downwind. None of the flies moved actively in the tradewinds typical of Kamuela.

C. *Analysis*

Dispersal over time is estimated by fitting a weighted regression. Since D_X and d_X are of form X_i/n, the arcsin root transformation is fitted to give for *D. mercatorum*:

$$Arc\ sin\ (\sqrt{D_X}) = .6029 - .0021X$$

The regression value .*0021* is not significantly different from zero, which means that the probability of leaving the rearing site is not significantly different from zero after the first day. All the dispersal occurred during KONA weather and 31% of the *D. mercatorum* left the cactus at that time. The average distance moved gives a similar result: $d_X = 42.71 + 5.3_X$. Again the regression value *5.3* is nonsignificant and no increase in dispersal was seen beyond the initial KONA weather. The average dispersing *D. mercatorum* flew only 43 meters and 31% of the population showed this short movement. *D. hydei* were considerably more dispersive than the *D. mercatorum*. The regression of proportion dispersed over time gave:

$$Arc\ sin\ (\sqrt{D_X}) = .7705 + .0686X$$

Again the regression was not significant and there was no significant dispersal during the trades. There was, however, a large proportion (68%) dispersing during KONA weather. The larger *D. hydei* dispersal is reflected also in distance moved:

$$d_X = 166.91 + 30.17X$$

Again, the regression is nonsignificant; *D. hydei* dispersed an average of 167 m during KONA weather but did not move significantly further after that.

Following Endler (1977, p. 60) we can find the root mean square dispersal distance as the product of the average

distance moved and the root of the probability of leaving.
This gives 24 meters for *D. mercatorum* and 138 meters for
D. hydei.

D. *Comparisons with Other Drosophila*

Comparing dispersal in *D. mercatorum* and *D. hydei* with
that of other *Drosophila*, we first note that the 6-7 fold
difference between *D. mercatorum* and *D. hydei* is the only
such difference for two *Drosophila* on a single substrate,
and reflects very different strategies, with *D. hydei* the
colonizer. Further, comparisons with other *Drosophila* show
that *D. mercatorum* is 1/2 to 1/1000 as dispersant as other
Drosophila studied (see review in Johnston and Heed, 1976;
Endler, 1979; and Jones *et al.*, 1981). This observation,
however, must be modified by habitat considerations. The
cacti at Kamuela are large and support continuous populations. We can convert the dispersal rate, therefore, (after
Endler, 1979), from meters into "average rot units" by
dividing root mean square dispersal rate by the average distance between suitable breeding sites. Since the area of
the study site is roughly 180 meters by 280 meters and
includes 26 cacti, the area per cactus is (180 X 280)/26 or
1936 meters squared. The square root of this gives the
average distance between cacti of 44 meters. This means that
the average distance moved by *D. mercatorum* is 24m/44m or .55
times the distance between cacti per dispersal day. This
value (.55) is equal to the value of .56 estimated by Endler
for *D. pseudoobscura* and its relatives and is comparable to
that found for *D. pachea, D. nigrospiracula, D. mettleri,*
D. mojavensis on giant cacti (Johnston and Endler, unpublished).

A comparable estimate for *D. hydei* gives 138m/44m or
3.1, which is larger than that for any other *Drosophila*
studied. Perhaps this reflects the fact, noted above, that
D. hydei are reared only from a subportion of the pads that
support *D. mercatorum.* The number of suitable breeding sites
for *D. hydei* may, therefore, be less than that used to estimate mean rot distance for *D. mercatorum,* and the mean distance between cacti may be an underestimate for *D. hydei.*

E.   Gene Flow

For genetic studies, the important character is not dispersal, but gene flow. Disperal means the average or root mean square dispersal distance, gene flow is the average egg to egg distance between succesive generations. The differences may be profound. A number of factors may alter rate of gene flow, including egg laying history (when and at what distance eggs are layed), survival (time to disperse), and mating structure, and in particular expected dispersal distance of male contributed alleles. In order to estimate gene flow, an attempt must be made to correct dispersal for each of these factors.

Use of laboratory strains to estimate egg laying schedules is illustrated in Endler (1979). He shows that a square root normal distribution gives a good fit to laboratory egg laying data and that the eggs are deposited at a distance which is 6/10 of the life time dispersal distance.

Survivorship data from recapture studies give the 1981 survival of *D. mercatorum* as .81 per day. Survival for *D. hydei* was .88 per day. For 1980 recaptures, the corresponding survival rates were .965 and .917 respectively (Templeton and Johnston, 1982). Direct age data (Johnston and Ellison, 1982; Figure 2) agrees well with these estimates. The survival rate estimates mean that less than 5% of the *D. mercatorum* survived more than two weeks in 1981 or more than four weeks in 1980. Variance in distance moved increases linearly with each added dispersal day, so root mean square dispersal should increase linearly with the root of dispersal time, and egg to egg distance for *D. mercatorum* should be the product of root mean square dispersal, egg laying, and survival, which gives:

$$l_c = 24m \ (0.6) \ \sqrt{2} = 20.4 \ m$$
$$l_c = 24m \ (0.6) \ \sqrt{4} = 28.8 \ m$$

for 1981 and 1980, respectively.

Mating structure is the final unknown and can be very important. If females mate only once with local males before dispersing and mate randomly with respect to age, then male and female contributions to gene flow are identical. If, however, the females mate with migrants, or after dispersing, or mate only with the oldest, most experienced males, then gene flow is limited to the female gametes. The mating structure is unknown for *D. mercatorum* and *D. hydei*,

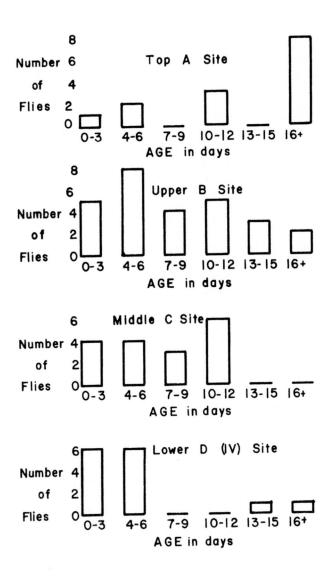

FIGURE 2. *Ages of field captured flies from sites A, B, C, D from 1980. The ages were determined after Johnston and Ellison (1982) by counting age layers formed on internal muscle attachments. The maximum reliable estimated age is 16 days, so the oldest flies are a minimum of 16 days, but may be considerably older.*

yet there is evidence of nonrandom mating of one or another
type since dispersal appears to be the same for males and females, and yet the "abnormal abdomen" character occurs at
different, negatively correlated frequencies in the X and Y
chromosomes of *D. mercatorum* (Templeton and Johnston, 1982).

F.   Neighborhood Size

Ideally gene flow is translated into neighborhood size,
or the size of a population which encompasses 95% of the egg
to egg transmission over one generation.   For *D. mercatorum*,
this is given as the number of individuals in an area whose
radius is root two times the mean square dispersal distance
(Wright, 1968; Endler, 1979).   For *D. mercatorum* in 1981,
this translates as an area which includes two cacti.   Given
the estimated population sizes for cacti in 1981 (Templeton
and Johnston, 1982) the effective neighborhood size of
D. mercatorum ranges between 260 flies for B-1 and B-2 and 11
flies for A-1 and A-2.   Given these populations, Wahlund
effects reported by Templeton and Johnston (1982) for 1981
are not unexpected.   Further, the large dispersal of *D. hydei*
and the greater population sizes of the 1980 *D. mercatorum*
make drift of much less consequence in *D. hydei* and
D. mercatorum   in 1980.

IV   Discussion and Conclusions

The species clines observed in *D. mercatorum* and *D. hydei*
reinforce the conclusions made here that *D. mercatorum* is
weakly dispersant, reluctant to move during high winds, and
rarely reach the most upwind sites.   At the same time, *D.
hydei* are relatively mobile and will move into the wind.
The absence of *D. hydei* at the base of the hill is not explained, however, by these observations, and remains a subject of speculation.   It is possible that the relatively
high fecundity and shorter generation time of *D. mercatorum*
results in competitive exclusion in the dry lower habitat.
Yet more rearing records and laboratory tests are needed to
confirm this supposition.
The difference in dispersal rates between *D. mercatorum*
and *D. hydei* coupled with very little difference in population size is unique and should provide an excellent opportunity to test population genetics theory, such as the hypothesis put forth by Slatkin (1981) that gene flow alone

determines the conditional average frequency of alleles in a
population.

Finally, the numerous theoretical papers which relate
population size, gene flow and selection in a cline let us
estimate selection on the cline in *D. mercatorum*. Templeton
(1980) and Templeton and Johnston (1982) suggest that the
cline in abnormal abdomen results from the greater intrinsic
rate of increase of the "aa" type at the lower site, while
longevity and competitive ability favors the "non-aa" at
the wetter and cooler upper sites. The extent of the selection required to produce the observed cline can be estimated
following Endler (1979) as

$$w = 2.45 \quad l/\sqrt{s}$$

where w is the width of the cline or 3000 meters, l is the
calculated gene flow or 28.8m, and s is the selection coefficient across the cline. Using these values, the selection
is calculated to be 5.8 times 10^{-4}.

This estimate depends upon the width of the cline being
equivalent to the distance between the uppermost and lower-
most collection sites and upon the cline height being the
result of genes which exhibit no dominance. One of us
(Templeton, unpublished) found a heterozygote advantage
for development time in certain combinations of "aa" and
"non-aa" lines, but the advantage was never large and
existed only for a few combinations of lines. Finally, Jain
and Bradshaw (1966) showed that different generation times
of the individuals on the two sides of a cline will have an
effect equivalent to an asymmetry in gene flow and will dis-
place the cline in the direction of the shorter generation
times. The cline may therefore be displaced somewhat
downhill toward the shorter generation times of the "aa"
types on the lower slope. This shift would mean that the
width of the cline is underestimated and that the selection
may be smaller than estimated. In contrast, any heterozygote
advantage results in a decrease in the cline height and an
underestimate of the true selection coefficients.

As a final conclusion, it must be noted that these results are in agreement with the field observations of clines
across short distances in the face of high amounts of dispersal in *D. aldrichi* (Richardson, 1969) and with the experimental results of Endler (1979) in which 40% dispersal
between local demes permitted the buildup of clines between
demes. Our results show that a cline in gene frequency can
be produced in the face of very high, but local dispersal
rates, if the population is subjected to levels of selection

on the order of five times 10^{-4}. Further, the cline can persist for years, until a change in the habitat produces a change in the selection regime, after which time the cline can rapidly disappear.

REFERENCES

Barton, N. (1979). *Heredity 43*, 341.
Endler, J. (1977). "Geographic Variation, Speciation, and Clines." Princeton University Press, New Jersey.
Endler, J.A. (1979). *Genetics 93*, 263.
Felsenstein, J. (1976). *A. Rev. Genet. 10*, 253.
Jain, S., and Bradshaw, A. (1966). *Heredity 21*, 407.
Johnston, J.S. (1978). *In* "The Screw-worm Problem" (R.H. Richardson, ed.), p. 151. University of Texas Press, Austin.
Johnston, J.S. (1982). *Behav. Genet.* (in press).
Johnston, J.S., and Ellison, J. (1982). *J. Insect Physiol.* (in press).
Johnston, J.S., and Heed, W.B. (1976). *Am. Nat. 110*, 629.
Jones, S., Bryant, S., Lewontin, R.C., Moore, J.A., and Prout, T. (1981). *Genetics 98*, 157.
Nagylaki, T., and Lucier, B. (1980). *Genetics 94*, 497.
Parsons, P.A., and McKenzie, J.A. (1972). *Evol. Biol. 5*, 87.
Richardson, R.H. (1969). *Jap. J. Genet. 44*, 172.
Richardson, R.H., and Johnston, J.S. (1975). *Oecologia 20*, 287.
Slatkin, M. (1981). *Genetics 99*, 323.
Taylor, C.E., and Powell, J.R. (1978). *Oecologia 37*, 69.
Templeton, A.R. (1980). *Theor. Populat. Biol. 18*, 279.
Templeton, A.R., and Johnston, J.S. (1982). Chapter 15, this Volume.
Wright, S. (1968). "Evolution and The Genetics of Populations," Vol. 4. University of Chicago Press, Chicago.

17
Adaptations to Competition in Cactus Breeding *Drosophila*

R. L. Mangan[1]

Department of Entomology
The Pennsylvania State University
and
U.S. Regional Pasture Research Laboratory
University Park, Pennsylvania

I Introduction

The results I present here are derived from a series of experiments designed to apply competition theory and laboratory simulation to the cactus-yeast-fly system. Ayala (1969a, 1969b) showed that for a two *Drosophila* species competition system a curvilinear or nonlinear adaptation of the Volterra competition model was required to explain the establishment of equilibrium. In a following series of experiments using vector analysis of a series of initial density combinations, Ayala *et al.* (1973) showed that an exponential parameter (Θ) associated with the density dependent parameters in the equation

$$\frac{dn_i}{dt} = r_i N_i \ (K_i^{\Theta} - N_i^{\Theta} - \frac{\alpha_{ij} N_j}{K_i^{1-\Theta}}) \ \frac{1}{K_i^{\Theta}}$$

provides a statistically real and biologically logical improvement of the equation. Ayala *et al.* (1973) suggested that the value of Θ, which in their system was approximately 1/2, is related to substrate characteristics.

---

[1]*Present address: USDA Screwworm Research, Mission, Texas.*

ECOLOGICAL GENETICS AND EVOLUTION
ISBN 0 12 078820 9

My original purpose in this series of experiments was to examine intra- and interspecific competition on natural substrates to derive and compare competition coefficients. Since we had a pretty good idea from field sampling and laboratory experiments (Fellows and Heed, 1972) what the "answer" should be, the experiments should have given a biologically real appraisal of the nonlinearity of the competition coefficients.

In carrying out these experiments, I had to make a series of assumptions and choices concerning interpretation of how the fly-cactus system works. Fortuitously, one of my major assumptions concerning survival and fitness proved wrong in the first experiment. Revisions of assumptions, analysis of experimental results, and collection of field data to test predictions from laboratory-derived results provided the information I will discuss here.

II   Components

Three *Drosophila* species, *D. nigrospiracula*, *D. mettleri*, and *D. mojavensis* are considered here as potential competitors. *Drosophila nigrospiracula* is excluded from the organpipe and agria substrates of *D. mojavensis* by chemical factors and probably from the soil substrate of *D. mettleri* by nutrient requirements (Heed, 1977). Experimental evidence suggested, therefore, that *D. nigrospiracula*'s saguaro-cardon niche is a subset of potential *D. mojavensis* niches. Since *D. mettleri* uses exudate from saguaro-cardon, its potential for use of the fresh tissue *D. nigrospiracula* niche is less certain. One goal of this research was to derive criteria independent of the nutrient competition tests to predict degree of niche overlap in cases where plant chemistry is not involved. I used mouthpart comparisons to make these predictions.

Cactus ecology was investigated in two major community types. Habitats dominated by saguaro and organpipe as *Drosophila* substrates were sampled in northern Sonora between Pitiquito and Libertad and on the Papago Indian reservation in Arizona. Our main censusing effort was in the agria-cardon habitats north of Bahia Kino on the gulf coast of Sonora. Substrate availability varies in both space and time for cactophilic diptera. For columnar cactus-adapted *repleta* group (*Drosophila*) species, locating larval and adult substrates certainly requires an adult phenotype

which is adapted to cactus substrates' nutritive and
distributive characteristics.

I chose the Kino area agria-cardon association for a
major censusing effort because both the cactus species and
the fly species seemed to show greatest contrast in this
area. Agria cactus grows in dense patches; seasonal

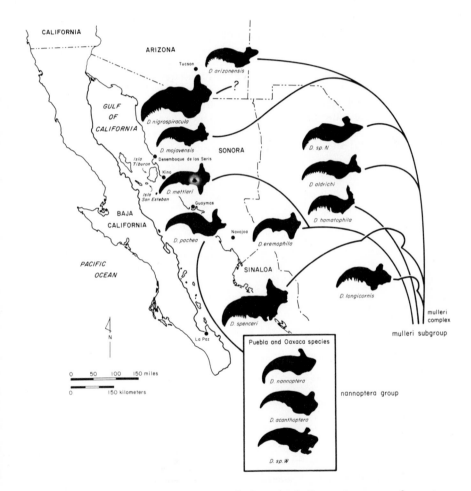

FIGURE 1.   *Mouthpart morphology of four Sonoran desert
endemic* Drosophila *spp. (farthest right) and their closest
cactus associated relatives. Phylogenetic interpretation
from Wasserman (pers. comm.).*

variation in water storage apparently causes frequent
crushing and necroses of stem tissue.   Cardon cacti in the
same area show a more uniform distribution and necroses are
associated with senescence or physical damage to the plant.
Drosophila mojavensis populations in this region are of the
smaller, chromosomally variable race, while the cardon-
adapted *D. nigrospiracula* population is chromosomally and
morphologically homogeneous throughout the Sonoran desert.
From less extensive data, *D. mettleri* also appears to be
homogeneous throughout the desert.

In laboratory experiments designed for relevance to a
real ecological community, I made independent observations
to collate and form a set of criteria for interpretation of
laboratory results.   The set of observations I used to bias
my experimental design included mouthpart structures
(Figure 1) and the thorax size–cactus size relationship
(Figure 2).   Mouthpart analysis suggested that *D.
nigrospiracula* and *D. mojavensis* should use more similar

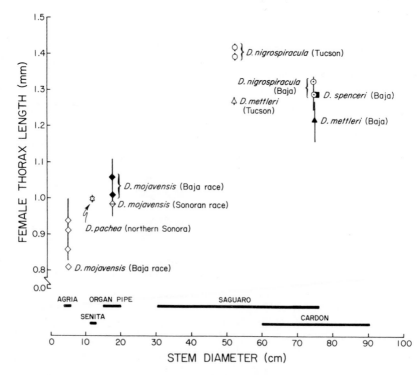

*FIGURE 2.   Relationship between female size and cactus
size.   Sites in parenthesis indicate collection site.*

resources than either would share with *D. mettleri*. A
second observation is that none of the four Sonoran desert
endemic species (*D. nigrospiracula, D. mettleri, D.
mojavensis,* and *D. pachea*) appears to have changed mouthpart
structure and, presumably, feeding habits from ancestral
forms as an adaptation to presently used substrates.

The relationship shown in Figure 2 indicates that
larger cactus stem diameter is associated with larger flies.
This could be due to larger flies being more fecund, larger
cacti producing more persistent rots giving flies longer to
develop, or larger cacti having more dispersed necroses.
Fecundity and thorax size are correlated within and among
the species in Figure 2. *Drosophila mettleri* used exudate
in soil, not the rot itself, suggesting that thorax size and
fecundity are not determined simply by substrate carrying
capacity. *Drosophila nigrospiracula* and *D. mettleri* are
shown to have similar thorax sizes in the same habitats.

III   Measurements of Competitive Fitness

The optimal phenotype for cactus rot utilization in the
Sonoran desert was evaluated in terms of three criteria:
1) The flies must be able to develop on the rot substrate
during a time period limited by the time the rot is
available.   2) Adults must be able to disperse sufficient
distances to colonize new rots.   3) Adult females must be
sufficiently fecund to maintain populations capable of
tracking resource fluctuations.   Physiological processes and
their degree of genetic control for adult size, fecundity,
and development rate have been investigated by Robertson
(1963, 1965), Bakker (1961), and Sang (1949a,b,c) for
laboratory populations of *Drosophila*.   Since development
rate, adult size, and correlated fecundity are
environmentally sensitive, these traits as well as mortality
seemed to be good fitness characters.

In a series of preliminary experiments in which I used
only survival percentages to estimate fitness, adult *D.
nigrospiracula* produced at densities leading to one species
being excluded were smaller than those ever collected or
reared from natural substrates.   I therefore suggest that
there may be a minimum effective adult size and a maximum
development time for population survival.   Optimal
phenotypes will depend on rot distribution, size, and
longevity.

IV   Experimental Approach

Like most *Drosophila* species, the *repleta* group has
life history characteristics fitting the colonizing or "r"
type profile.   The major questions I will address here are:
1) How do the three species respond to degenerating larval
substrates due to microorganismic metabolic processes and
nutrient consumption by *Drosophila* populations?   2) How are
these responses correlated with the primary larval habitats
of the populations?

Three aspects of fitness were investigated.   Egg to
adult survival, development time, and thorax size were
chosen to comprise the fitness set for these species.
Survival is related to behavioral response to experimental
conditions, particularly pupation site selection and larval
foraging.   Development time is a primary component of
colonizing fitness especially in ephemeral habitats.   Thorax
size is correlated with fecundity as well as migratory
ability (Roff, 1977).

The experimental protocol used for competition
experiments is shown in Figure 3.   Flies and saguaro cactus
tissue were collected in the field.   *Drosophila mojavensis*

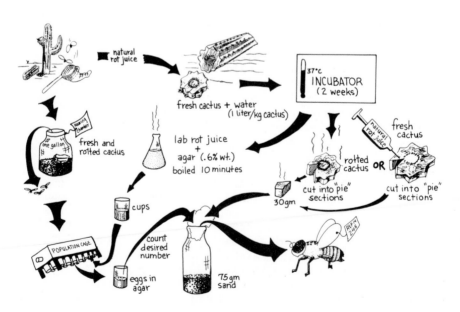

*FIGURE 3.   Schematic representation of competition
experiment design for competition on Saguaro cactus tissue.*

were collected on the Papago reservation or from Pitiquito, Sonora; *D. nigrospiracula* and *D. mettleri* were collected from those sites or the Tucson vicinity.

Populations were expanded to several hundred flies in one-gallon jars with approximately 700 g of rot exudate inoculated cactus. Adults were placed in population cages with banana food for adult feeding and saguaro-necrosis juice and agar gel for oviposition.

Half-pint milk bottles were used as competition chambers and two types of substrates were used. Substrates designated as "old" were prepared by cutting a 10 kg saguaro trunk into a number of 3 cm slices. These pieces were placed loosely in one-gallon jars and covered with warm water. Approximately 100 ml of gauze-filtered natural rot exudate were collected in the field and added to the jars, then incubated at 37°C for 14 days. Pieces of this cactus (30 g) were added to bottles containing 75 g of sterile sand and stored in a freezer (-20°C) until use in experiments.

Cactus tissue designated as "fresh" was prepared each day that chambers were initiated. The saguaro limbs were cut into slices (3 cm) and cortex was cut into 30 g cubes. These cubes were inoculated with 1 ml of filtered natural rot exudate and placed in bottles with 75 g sterile sand.

As eggs of both species were available, old tissue bottles were thawed or fresh tissue bottles prepared. Eggs of both species were introduced to bottles at the same time. Usually oviposition cages were prepared in the evening and checked the following morning. The maximum age difference in eggs was never greater than 12 to 16 hours.

Eggs of *D. nigrospiracula* were combined with those of *D. mojavensis* or *D. mettleri* in a series of tests. For both pairs of species, 25 combinations of eggs were tested on fresh tissue and on old tissue. Densities of 30, 50, 80, 120, and 170 eggs of both species of each pair were tested in a five-by-five matrix of all combinations. All bottles containing 30 eggs of either species were replicated twice. Thus, for each pair, *D. nigrospiracula* vs. *D. mojavensis* or *D. nigrospiracula* vs. *D. mettleri*, 34 bottles were set up on two substrate types, fresh and old. Approximately 10 bottles were initiated per week.

Bottles with substrate and eggs were placed in a Percival Incubator I-30B. Temperature and lights were controlled on a 12 hr dark (16°C) and 12 hr light (27°C) cycle. Checks of egg viability showed *D. mojavensis* and *D. nigrospiracula* typically had 100% egg eclosure; *D. mettleri* eggs frequently had as high as 50% noneclosure. Viability of *D. mettleri* eggs was increased to about 95% by using older (15 days) females in cages with an excess of males and by excluding all eggs with nurse cell remnants (transparent areas) at the filament ends. Failure of eggs to hatch resulted in six bottles being omitted in the *D. mettleri* vs. *D. nigrospiracula*, fresh cactus experiment where large numbers of *D. mettleri* eggs were required.

Bottles were checked daily for emergences. Species, sex, and date were recorded for each emerging fly. All females from bottles with 30 eggs of either species were saved in *Drosophila* food vials for 5 days, then thorax measurements were recorded.

V   Results — Competition Chamber Study

For survival analysis, the statistical test of interest is the ANOVA for density effects on survival (Table I). Mortality in this test is largely due to larvae migrating into the cotton stoppers of the vials. In all cases with significant effects, there was declining survival at higher densities.

More sensitive regression analysis could be performed with thorax length and time for development data. Here, for the two species i and j, linear regressions relating size or development time (Y) with egg density of either of the two species were analyzed in the models

$$Y_i = \alpha + \beta x_i$$

$$Y_i = \alpha + \beta x_j$$

and

$$Y_i = \alpha + \beta x_i + \beta' x_i^2$$

$$Y_i = \alpha + \beta x_j + \beta' x_j^2$$

using F test and r^2 values as criteria to evaluate the best models. In each pair of equations, the first evaluates intraspecific effects while the second evaluates interspecific effects. A significant improvement due to addition of the $\beta' x^2$ term indicates a curvilinear relationship between larval density and size or development time.

Shapes of curves typical for thorax size-egg density and development time-egg density relationships are shown in

TABLE I.   ANOVA of Survival Percentages

| Interaction | | Old tissue | | Fresh tissue | |
Dependent	Independent	$F_{16,4}$	Signif.	$F_{16,4}$	Signif.
D. mojavensis	D. mojavensis	4.05	0.02	8.14	0.00
	D. nigrospiracula	4.76	0.01	2.99	0.05
D. nigrospiracula	D. nigrospiracula	4.64	0.01	0.31	ns
	D. mojavensis	1.05	ns	0.29	ns
D. mettleri	D. mettleri	6.24	0.00	*	*
	D. nigrospiracula	2.75	ns	*	*
D. nigrospiracula	D. nigrospiracula	1.15	ns	*	*
	D. mettleri	3.24	0.03	*	*

* *Not significant* - $F_{3,9}$

Figure 4.   In this figure, egg density increase results in a loss of fitness. For thorax size, a negative second derivative implies fitness loss at an increasing rate, and a positive second derivative implies fitness loss at a decreasing rate. A decreasing rate of stunting is usually associated with mortality or increased development time. For development time, the interpretations are the opposite since increased development time results in decrease in population growth rate and an increase in probability of substrate degeneration during the larval stage.

Differing patterns in changes in size (Table II) on the two tissue types suggest differences in resource use. *Drosophila nigrospiracula* has positive second derivative interaction for intraspecific interactions on both tissue types but responds to D. *mettleri* and D. *mojavensis* differently on different tissues. *Drosophila mettleri* responds to intraspecific density in the same manner on the two tissue types with a negative second derivative.

Differences in interspecific interactions may be due to dissimilarity of resource use among these species. For every pair of species, the second derivatives are dissimilar on fresh and old cactus. Since adult size is determined during the feeding period just before pupation (Robertson, 1965), which should be the period of greatest competitive stress, it appears that the two substrate types are being used differently by each species. Since the second derivative is a measure of the curvilinearity of the interaction, differences in the sign of this parameter

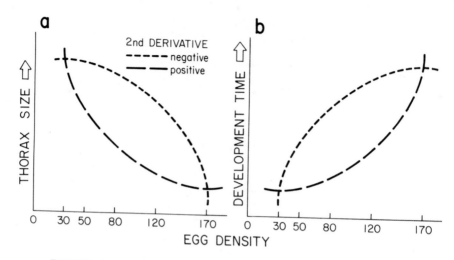

FIGURE 4.   *General shapes of curves relating fitness (thorax size (a), development time (b)) to egg density with positive or negative 2nd derivatives of quadratic model.*

indicate that nonlinear parameters in competition equations are sensitive to substrate conditions as proposed by Ayala *et al.* (1973).

Development time (Table III) response curves may be interpreted in a manner similar to thorax size curves. Slowly developing *D. mettleri* response curves all have a positive second derivative. Since this species uses exudate soaked soil under natural conditions, response to cactus age and *D. nigrospiracula* competition here is probably a result of changes in the quality of the exudate in the sand. Responses of *D. mojavensis* to increasing density of itself and *D. nigrospiracula* are much weaker on both tissue types. Probably the best interpretation for *D. mojavensis* development rate data is that while older tissue causes a decrease in rate, this species does not respond to competition by slowed development. *Drosophila nigrospiracula* responds to interspecific competition with *D. mettleri* and *D. mojavensis* in different manners. On fresh tissue the *D. nigrospiracula* response to *D. mojavensis* has the highest r^2 observed. The response to *D. mettleri* is not significant. On old tissue, shapes of response curves are reversed for the two species. *Drosophila nigrospiracula* also responds to its own density differently on the two tissue types.

TABLE II. Summary of Regression Analyses of Quadratic Models Relating Thorax Length to Egg Density (2nd DER is second derivative of model shown graphically in Figure 4.)

Dependent	Treatment	Female thorax size				
		Tissue	2nd DER	r^2	$\bar{x}$(mm)	CV%
D. nigrospiracula	D. nigrospiracula	Old	+	0.58	1.09	9.5
		Fresh	+	0.41	1.38	4.7
D. nigrospiracula	D. mojavensis	Old	+	0.40	1.20	6.4
		Fresh	0	0.04*	1.46	2.6
D. mojavensis	D. mojavensis	Old	+	0.48	0.87	8.8
		Fresh	−	0.54	1.06	5.7
D. mojavensis	D. nigrospiracula	Old	+	0.27	0.92	7.0
		Fresh	−	0.74	1.04	6.0
D. nigrospiracula	D. nigrospiracula	Old	+	0.30	1.19	8.1
		Fresh	+	0.66	1.31	6.2
D. nigrospiracula	D. mettleri	Old	0	0.05*	1.28	8.3
		Fresh	−	0.61	1.45	3.5
D. mettleri	D. mettleri	Old	−	0.59	0.99	10.5
		Fresh	−	0.34	1.22	5.1
D. mettleri	D. nigrospiracula	Old	0	0.02*	1.01	9.0
		Fresh	+	0.34	1.23	4.7

* Not significant - P > 0.01

TABLE III. Summary of Regression Analyses of Quadratic Models Relating Development Time to Egg Density (2nd DER is second derivative of model shown graphically in Figure 4.)

Dependent	Treatment	Development time				
		Tissue	2nd DER	r^2	$\bar{x}$Days	CV%
D. nigrospiracula	D. nigrospiracula	Old	+	0.45	22.42	12.2
		Fresh	−	0.47	17.13	8.8
D. nigrospiracula	D. mojavensis	Old	−	0.42	22.19	10.7
		Fresh	−	0.71	16.35	7.3
D. mojavensis	D. mojavensis	Old	−	0.30	18.97	9.1
		Fresh	−	0.19	17.02	12.0
D. mojavensis	D. nigrospiracula	Old	−	0.25	19.59	9.6
		Fresh	+	0.21	17.04	9.1
D. nigrospiracula	D. nigrospiracula	Old	+	0.09*	16.05	11.2
		Fresh	−	0.29	16.31	9.2
D. nigrospiracula	D. mettleri	Old	+	0.33	16.80	7.4
		Fresh	+	0.07*	15.12	4.8
D. mettleri	D. mettleri	Old	+	0.61	18.20	10.3
		Fresh	+	0.63	21.09	12.1
D. mettleri	D. nigrospiracula	Old	+	0.18	20.55	8.8
		Fresh	+	0.58	20.00	11.5

* Not significant - P > 0.01

VI   Discussion

Substrates in these bottles provided heterogeneous resources for the *Drosophila* larvae. Larvae faced with a shortage of nutrients could respond in several ways. Responses measured here included size reduction, slowed development, and death, a probable consequence of migration. Curvilinear responses are interpreted as responses to the disappearance of these nutrients. If the order of resource preference of two *Drosophila* species differs, the shapes of their interspecific and intraspecific response curves will be different. These results suggest that the resource preferences of the three species are different and the resource sets offered by fresh and old substrates differ. I do not know what these resources are, but they may be related to different microorganism species or products or different stages of cactus necrosis.

The *Drosophila* species have several adaptive "choices" in response to resource depletion. I have summarized the responses in these experiments in Table IV. For a species to be considered to respond by mortality, the ANOVA F test was used. For size and time responses, there had to be a minimum degree of variation (CV > 5%) and coefficient of determination (r^2 > 0.25).

On old cactus all three species respond to increases in their own density by all three criteria. *Drosophila nigrospiracula* responds to *D. mojavensis* by size reduction and slower development and to *D. mettleri* by mortality and slower development. Response of *D. mojavensis* to *D. nigrospiracula* is by mortality and stunting, but *D. mettleri* shows no significant response to *D. nigrospiracula*. The major conclusion here is that *D. mettleri* appears best adapted to competing on depleted tissue.

On fresh tissue it is interesting that *D. mojavensis* responds only by mortality or size reduction. *Drosophila nigrospiracula* only responds by time or size but not mortality. When interspecific effects are examined, it is notable that *D. nigrospiracula*'s competitors lose fitness for traits in which they differ most from *D. nigrospiracula* under increasing interspecific competition. *Drosophila mojavensis* gets smaller while *D. mettleri* develops more slowly. I find it satisfying that the saguaro rot inhabitant, which may be assumed to have the optimal phenotype for this environment, induces competitors to change their phenotypes in traits for which they are already most different from this optimum.

TABLE IV. *Summary of Competition Effects (Criteria are given in text.)*

Responding species	Increasing species		
	D. mojavensis	*D. nigrospiracula*	*D. mettleri*
Old cactus			
D. mojavensis	mortality/size/time	mortality/size	no test
D. nigrospiracula	size/time	mortality/size/time	mortality/time
D. mettleri	no test		mortality/size/time
Fresh cactus			
D. mojavensis	mortality/size	mortality/size	no test
D. nigrospiracula	time	size/time	
D. mettleri	no test	time	size/time

VII Cactus Distribution and Substrate Availability

In evaluating the components of fitness, I showed in Figure 2 the relationship between thorax size and cactus stem size for five cactus associated *Drosophila* in Sonora and Baja, California. Data in Tables II and III indicate that the smallest fly, *D. mojavensis* gets still smaller under nutrient stress, and the slowest developing fly, *D. mettleri*, develops even more slowly. Since thorax size is related to migratory and foraging ability (Roff, 1977), data in Table V give a comparative survey of the foraging requirements of these species.

Cacti for the agria subrace of *D. mojavensis* are the most dense in this area. Comparing densities of cacti shows, that in terms of cacti per hectare, agria at a density of 13.6 cacti is twice as common as cardon (6.8 per hectare) and about 7 times more common than organpipe (1.9 per hectare). Flies can only use necrotic cacti, however, and densities of necroses are more variable among cactus species. Agria rots (1.579 per hectare) are about 100 times more frequent in these areas than cardon (0.016 per hectare). While organpipe rots (0.040 per hectare) are comparable to cardon in spatial density, organpipe rots per cactus (0.021) probability is about midway between agria (0.116 rots per cactus) and cardon (0.002 rots per cactus).

TABLE V.   *Cactus and Necroses Numbers from Agria Habitats on Sonora Coast (Onah, Arenas, Desemboque) and Gulf Islands (Esteban, Tiburon)*

Sample	Area $x(100m^2)$	Organpipe Cacti	Organpipe Rots	Senita Cacti	Senita Rots	Agria Cacti	Agria Rots	Cardon Cacti	Cardon Rots
Onah November	4770	126	5	36	2	432	94	209	2
Arenas November	1500	46	0	0	0	115	16	172	0
Onah February	4710	139	1	68	5	597	112	242	0
Desemboque February	4200	7	1	17	0	542	74	550	1
Onah May	4800	123	3	40	0	460	52	181	1
Esteban May	2250	1	0	0	0	661	26	358	0
Tiburon May	2970	35	0	0	0	613	24	3	0
Total	25200	477	10	161	7	3420	398	1715	4

Since *D. mojavensis* can use all three of these cacti, the actual density of substrates for this species is 0.05 rots per hectare higher than the agria rot density over all the sampling dates.   Comparison of November–February samples with May (dry season) samples at the Onah site suggests that agria rots may be more seasonal than organpipe and cardon, a factor that may select for *D. mojavensis* use of these resources.

The major conclusion I draw from the cactus and rot census is that in the agria regions of Sonora and the gulf islands, cactus rot density and the probability of rot per cactus correlate with the size of *Drosophila* adults.   Substrate characteristics, including predictability, have been shown in studies of Hawaiian *Drosophila* (Kambysellis and Heed, 1971; Montague *et al.*, 1981) and temperate domestic *Drosophila* (Atkinson, 1979) to be important factors in reproductive effort.   From Figure 2 and Table V data, cactus size and substrate distribution cannot be separated as factors selecting for adult size.

VIII Conclusions

The relationships between fitness components, habitat characteristics, and realized niches of three *Drosophila* species have been presented. Generally I have argued that responses to competition agree fairly well with both observed patterns of substrate utilization and with larval mouthpart structure. Patterns in nonlinear parameters in the expanded Volterra equation were shown to be sensitive to both substrate types and to the contestants. I interpret these patterns as indications that the substrates tested, old and fresh cactus, present a different series of nutrient resources which are preferred in different order by the three species. Curvilinearities in competition parameters are now being accepted or at least assumed in theoretical work (e.g., Rosenzweig, 1981). Density or frequency dependent changes in competition coefficients appear to me to be another manifestation of the problem of identifying the common resource for which two or more populations compete. In the cactus–*Drosophila* case, number of resources, order of preferences, and amount of each resource will affect the shape of response curves.

Interpretations of how each species responds to competition are made relative to ecology of the natural preferred substrates. Host plant specificity and regional restriction of *D. mojavensis* can be explained in terms of substrate distribution; organpipe and agria cactus rots are more frequent and, in regions where these species are dense, are more likely to have necroses. I interpret the role of *D. nigrospiracula* in this restriction mainly as further reducing the colonizing fitness of *D. mojavensis* in areas where organpipe or agria are not present. Substrate specificity of *D. mettleri* may be determined by its mouthparts alone. In rot pockets in the desert, a variety of Diptera (Syrphidae, Neriidae) use older rot tissue (Mangan, 1982). Competition may, therefore, be a factor in habitat selection by *D. mettleri*.

REFERENCES

Atkinson, W. (1979). *J. Anim. Ecol. 48,* 53.
Ayala, F.J. (1969a). *Nature, Lond. 224,* 1076.
Ayala, F.J. (1969b). *Genet. Res. 14,* 95.

Ayala, F.J., Gilpin, M.E., and Ehrenfeld, J.G. (1973). *Theor. Populat. Biol. 4*, 331.

Bakker, K. (1961). *Arch. Néerl. Zool. 14*, 200.

Fellows, D., and Heed, W.B. (1972). *Ecology 53*, 850.

Heed, W.B. (1977). *Proc. ent. Soc. Wash. 79*, 649.

Kambysellis, M.P., Heed, W.B. (1971). *Am. Nat. 185*, 31.

Mangan, R. (1982). *Pan-Pacif. Ent.* (in review).

Montague, J., Mangan, R., and Starmer, W. (1981). *Am. Nat. 118*, 865.

Robertson, F.W. (1963). *Genet. Res. 4*, 74.

Robertson, F.W. (1965). *In* "The Genetics of Colonizing Species" (H.G. Baker and G.L. Stebbins, eds.), p. 95. Academic Press, New York.

Roff, D. (1977). *J. Anim. Ecol. 46*, 443.

Rosenzweig, M.L. (1981). *Ecology 62*, 327.

Sang, J.H. (1949a). *Physiol. Zool. 22*, 183.

Sang, J.H. (1949b). *Physiol. Zool. 22*, 210.

Sang, J.H. (1949c). *Physiol. Zool. 22*, 202.

18
Mating Systems of Cactophilic *Drosophila*

Therese Ann Markow

Department of Zoology
Arizona State University
Tempe, Arizona

The diversity observed in insect mating systems is of
considerable interest to behavioral ecologists and geneti-
cists (Blum and Blum, 1979; Krebs and Davies, 1978).   In
order to explain the forces which shape mating systems we
need to understand the long term phylogenetic relationships
and the ecology of the species being studied.   Evolutionary
relationships within the genus *Drosophila* have been extensively reconstructed (Patterson and Stone, 1952; Wheeler,
1981; Wasserman, 1982) and the ecology of a number of species,
especially the cactophilics, has been the subject of intensive studies (Heed, 1978).   Cactophilic species of *Drosophila*
are predominantly from the very large *repleta* species group
and the *nannoptera* species group.   Information about their
host plants, geographic ranges and nutritional requirements
is contained in the chapters by Heed, Starmer, and Kircher in
this volume.
  The mating systems of cactophilic *Drosophila* species have
recently been under investigation in my laboratory and have
been found to differ in major ways from the mating systems of
non-cactophilic *Drosophila*.   The nature of these differences
and the question of the evolutionary and ecological forces
which may have created them are the topics to be discussed
here.   First, the mating systems of a number of cactophilic
Drosophila species will be described.   The species and their
collection data are listed in Table I.   Special emphasis will
be placed on *D. mojavensis* for which the largest amount of
experimental data is available.   Then we will evaluate how
the mating systems of cactophilic *Drosophila* differ from
Drosophila species which utilize other types of resources,

ECOLOGICAL GENETICS AND EVOLUTION
ISBN 0 12 078820 9

TABLE I.   *Species Studied, and Location and Date of Collection of Strains Used in Studies of Mating Systems*

Species	Collection locality	Date
D. nigrospiracula	Santa Rosa Mountains, Arizona	2/81
D. mettleri	Santa Rosa Mountains, Arizona	2/81
D. mojavensis	Santa Rosa Mountains, Arizona	2/80
D. pachea	Hermosillo, Sonora, Mexico	1/69
D. arizonensis	Tempe, Arizona	4/79
D. hydei	Tempe, Arizona	6/81
D. nannoptera	Zapotitlan, Puebla, Mexico	7/70

and suggest how these differences might be related to their cactophilic existence. Finally, the relationship between the mating systems and the ecology of various cactophilic species will be explored.

I   Mating Systems of Cactophilic *Drosophila*

A. *Age at Reproductive Maturity*

In the course of a series of mating experiments performed in my laboratory and the laboratory of Dr. William Heed at the University of Arizona, it appeared that many cactophilic *Drosophila* species exhibit sexual dimorphism for the age at which reproductive maturity is reached (Cooper, 1964; Jefferson, 1977; Markow, 1981). Subsequently, this was measured for *D. mojavensis* in the laboratory (Table II). Ninety-seven percent of females were inseminated at three days of age, while many males did not inseminate females until they reached 8-10 days of age. Behavioural observations revealed that until males are about five days old they are not particularly interested in courting females and that young males who do court females do so much less persistently than do older males. These observations on sexual dimorphism in maturation have been generalized to include *D. nigrospiracula*, *D. mettleri* and *D. pachea* (Markow, Fogleman and Heed, unpublished).

TABLE II. *Age at Reproductive Maturity for Males and Females of D. mojavensis*

Sex	Age	Number mated	%
Males			
	3 days	43/156	27.6
	4 days	38/76	50.0
	5 days	62/82	75.6
	6 days	49/65	75.3
	7 days	63/77	81.8
	8 days	55/64	85.9
	9 days	74/81	91.3
	10 days	96/99	96.9
Females			
	1 day	2/47	4.2
	2 days	39/55	70.9
	3 days	58/60	96.7
	4 days	58/59	98.3

B. *Incidence of Female Remating*

A survey has been made of the frequency of female remating in a number of cactophilic *Drosophila* from the *nannoptera* and *repleta* species groups. These experiments were conducted in the laboratory by storing mated females apart from males and allowing them an opportunity to remate during a two hour observation period every morning for five days. The number of days required for females of each species to remate is shown in Figure 1(a). Females of *D. nigrospiracula* and *D. hydei* showed the shortest time until remating: nearly all females of these two species remated 24 hours after the first mating. In *D. nannoptera* some females required 48 hours to elapse before they would remate. The longest remating intervals were found for *D. pachea* and *D. mettleri* females. For comparison, similar data are presented for *D. melanogaster* females tested by the same procedures as employed with the cactophilic species.

In Figure 1(b), the proportion of females remating daily is presented. In *D. nigrospiracula* and *D. hydei* effectively all of the females remated daily. Approximately half of the *D. mojavensis* and *D. arizonensis* females remated daily. Daily remating was least frequent among females of *D. pachea*.

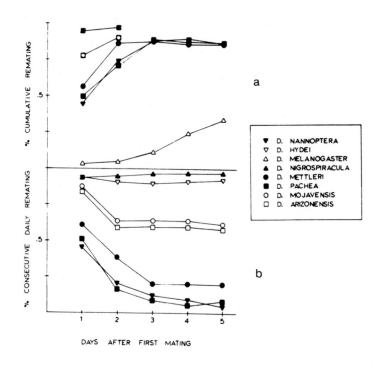

FIGURE 1.   (a) The time until females of cactophilic species remate the first time.   (b) Proportion of females remating daily for five days.

Since most *D. nigrospiracula* and *D. hydei* females remated at the 24 hour test, we determined how frequently they would remate if provided unlimited opportunities. Once females had mated they were transferred immediately to vials containing a new pair of virgin males and observed until they mated again. The number of times each female remated during a four hour observation period (8:00 a.m. – 12:00 noon) was scored. For *D. nigrospiracula*, the average number of matings was 3.89 ± 0.67 (n = 41) with several females mating five times in four hours. In *D. hydei* the average was 2.55 ± 0.94 (n = 38) matings. One *D. hydei* female mated four times. When females of either species are dissected after multiple matings, the ventral receptacles contain a number of discrete clusters of sperm equal to the observed number of matings. Within 12 hours, the sperm from all matings appear to be mixed within the receptacle. The influence of multiple matings in one day on daily mating incidence was investigated in

D. nigrospiracula, the species showing the highest frequency
of female remating. Females were mated twice on the first
day and then the number remating twice on consecutive days
was examined. On the first day after mating twice, 87%
(27/31) of the females remated two more times. On the next
day 80% (24/30, one female escaped) remated twice. Three
days later 77% (23/30) remated twice and on the last day 60%
(18/30) still engaged in a double mating. None of the other
species exhibited female remating during the same morning,
with the exception of *D. mettleri* where three different
females (out of 81) were observed to remate during the observation period.

*C. Effect of Remating on Female Fitness - Laboratory
Investigations with D. mojavensis*

Most *D. mojavensis* females remate a second time 24 hours
after their first copulation, as described above. The remainder remate at 48 hours when given a second opportunity.
Figure 2 shows the daily productivity rates for two weeks for
females that mated once, twice (remated after 24 hours), and
twice (remated after 48 hours) and females continuously
paired with males. In females that only mated once, productivity fell dramatically from the fourth day. Females which
mated twice laid eggs longer when the second mating was 48
hours after the first. The continuously paired females maintained their productivity the longest. That this is not
merely a function of the presence of males has been established by experiments in which males were present but
separated from females by a mesh partition. In terms of
total productivity, females paired continually produce the
most offspring (Mean offspring number $\bar{x}$ = 380). A second
mating did not increase overall productivity unless it
occurred 48 hours after the first ($\bar{x}$ = 82). Single mated
females gave the same number of offspring ($\bar{x}$ = 51) as twice
mated females who remated after only 24 hours ($\bar{x}$ = 53). Inspection of the ventral receptacles two weeks after a single
mating reveals large quantities of motile sperm. That the
viability of these sperm is still high is suggested by the
fact that egg to adult survival does not decline with time
after mating. Out of 100 eggs deposited by females one day
after mating 91 adult flies were obtained. Eggs collected
from single mated females ten days after mating had an adult
survival rate of 92%.

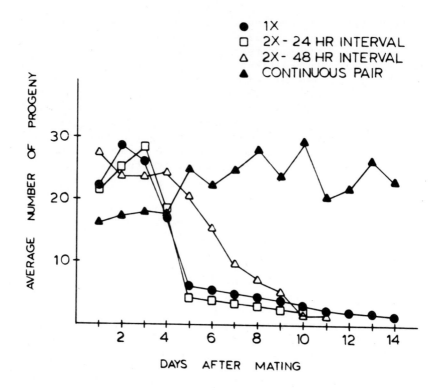

FIGURE 2. *Daily productivity (adult progeny) for D. mojavensis females for two weeks. Open squares = inseminated once, solid circles = inseminated twice 24 hours apart, open triangles = inseminated twice 48 hours apart, closed triangles = continuously housed with males.*

D. Effect of Remating on Male Fitness - Laboratory Studies with D. mojavensis

Two aspects of female remating for male reproductive bio-logy were considered. In *D. melanogaster*, males exhibit a dramatic reduction in fertility following the third succes-sive mating. Due to the overall increase in mating frequency by cactophilic species, we became curious about the fertility of multiply-mated male *D. mojavensis*. Males were presented with virgin females until they had mated seven times in suc-cession during a two hour observation period. Males showed no decrease in mating propensity over the seven matings. Females were saved and their offspring counted. The

offspring number did not decrease by the seventh mating
(Fig. 3). How *D. mojavensis* males maintain their fertility
level across so many matings might become clear from study of
factors such as quantity of sperm and the nature of ejaculate
components transferred at each mating.

The other problem that multiple female mating poses for
males is inter-ejaculate competition. Since females store
sperm, intrasexual selection can occur within the female reproductive tract. We were able to study the sperm utilization pattern in *D. mojavensis* by employing a recessive mutation causing dark eye color which arose spontaneously in the
laboratory. Dark-eyed females were mated either to dark-
eyed males first and then to wild-type males or to wild-type

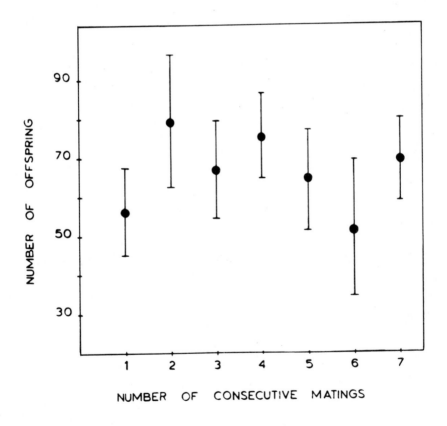

FIGURE 3. *Average productivity of males on their first
through seventh consecutive matings.*

males first and then to dark-eyed males. The relative number of dark-eyed and wild-type progeny were then used in calculating the proportion of offspring sired by the second male (P_2). The second mating occurred either 24 or 48 hours after the first. The relative contributions of the first and second mating are shown in Table III. A 2 X 2 factorial analysis of variance showed significant differences between the 24 and 48 hour time intervals. There was no effect of genotype of the second male or significant interaction between remating interval and genotype (Table IV).

While early remating may allow females to maintain their oviposition rates, the temporal change in P_2 values suggests an advantage exists for males that are able to assure that longer intervals have elapsed before they mate with a non-virgin. Such assurance would be afforded to males that are able to assess the time since a female last mated and in

TABLE III. *P_2 Values in Experiments where Rematings were 24 hours Apart and 48 hours Apart. The Genotypes of the First and Second Mates were Alternated for each Remating Interval*

Remating interval	Genotype of second male	P_2 (%)
24 hours	wild type	64.85 ± 2.48
	dark eye	68.17 ± 1.72
48 hours	wild type	97.63 ± 3.73
	dark eye	96.91 ± 4.78

TABLE IV. *A 2 X 2 Factorial Analysis of Variance on Arc-sin Transformed P_2 Values*

Source of variation	D.f.	Sum of squares	Mean squares	F
Total	15	5373.9		
Genotype	1	3.7	3.7	0.48
Time interval	1	5263.9	5263.9	684.51[a]
Genotype x time	1	14.1	14.1	1.83
Error	12	92.2	7.7	

[a] $P < 0.001$

doing so, increase the prospective success of their own sperm prior to investing time and energy in courtship. A series of male courtship choice experiments (Markow, 1981) was conducted to examine the possibility of pre-courtship discrimination. Individual virgin males, placed in a chamber with a virgin female and a mated female were observed until they courted a female. The proportion of the males that first courted a virgin instead of a non-virgin is seen in Table V. The data clearly show that males discriminate between females prior to investing in courtship. The preference for courting a virgin female does not disappear until the time corresponding to the male's ability to fertilize over 98% of subsequent eggs.

At the time that sperm utilization in *D. mojavensis* was investigated, the incredibly high remating frequency in female *D. nigrospiracula* and *D. hydei* had not been discovered. Obviously these species, especially *D. nigrospiracula* with its unique male-male interactions, present an interesting problem for paternity assurance and their sperm utilization patterns need to be investigated.

E. Inbreeding and Sterility in D. mojavensis

A potential relationship between remating and inbreeding avoidance is suggested by the outcomes of pair matings between related and between unrelated individuals. A variety of studies have required that pair matings be done for various desert species of *Drosophila*. These experiments frequently encountered difficulties because sibmated lines often failed to reproduce. An analysis of a similar situation in *D. melanogaster* showed that inbred lines failed because males are not stimulated to court closely related females (Averhoff and Richardson, 1974). The possibility that loss

TABLE V. *The Outcome of Male Choice Experiments where Single Males were Offered Two Females, One Virgin and One Having Recently Mated*

Interval since mated female had mated with other male	First female courted	
	Mated	Virgin
2 hours	2	46
24 hours	6	39
48 hours	18	28
96 hours	29	26

of fertility in *D. mojavensis* was occurring by a similar
mechanism was examined in inbreeding experiments in which
flies were sibmated.

A total of 22 sibmated lines was started, each from the
descendents of individual wild caught females. In the second
generation 10 pairs of progeny from a single female were
selected and placed in vials. The number producing progeny
out of the 10 pairs for each of the 22 lines was recorded.
Each subsequent generation was begun by taking 10 pairs from
one of the vials of a given line. This inbreeding scheme was
carried out for five generations. At the second generation,
seven of the 22 lines failed to reproduce. In each case, all
10 vials from a non-reproductive line were affected. A complete absence of eggs suggested that females be dissected and
examined for evidence of insemination. Five of the 10
females from each of seven lines were dissected, and the ventral receptacles of each female were found to be filled with
motile sperm. The other five females were then mated to unrelated males from other non-reproductive lines. All
females but one subsequently produced viable offspring
indicating that both males and females were fully fertile
with unrelated mates.

In each subsequent generation, about half of the inbred
lines became non-reproductive. In every case, all 10 of the
pairs failed to oviposit, although females did contain motile
sperm and both males and females could reproduce when re-
paired with unrelated individuals from other lines. The only
apparent deterrent to oviposition was the degree of relatedness of the pair. A factor or factors controlling reproduction may be segregating.

Males of *D. melanogaster* transfer an ejaculate component
which is required to stimulate oviposition (Garcia-Bellido,
1964; Merle, 1968). The observations in *D. mojavensis* could
be explained by the inability of such a component to trigger
egg laying when the male is too closely related to the
female, i.e., a sort of "self sterility" system. If such a
system were operating it would tend to favor outbreeding and
the maintenance of genetic variation in a population.

II Relationship of Mating Systems to the Ecology of Cactophilic *Drosophila*

In comparison with other species groups, the cactophilic
Drosophila and their relatives in the *repleta* and *nannoptera*
groups are characterized by sexual dimorphism for age at
reproductive maturity and frequent female remating. In

non-cactophilic species such as *D. pseudoobscura* (Pruzan, 1976), *D. simulans* (Markow, unpublished) and *D. melanogaster* (Pyle and Gromko, 1978) where females eventually remate, the average remating latency is a number of days. While the cactophilic *Drosophila* clearly remate more frequently than other species, considerable variability exists among them for this behavior. Thus two questions can be asked. First, why do frequent female remating and delayed male maturity occur in the cactophilic species? Second, what factors control the variability in remating frequency among these species?

In answer to the first question, the relatives of cactophilic species also exhibit female remating. Therefore it is possible that this trait existed among ancestral *repleta* and *nannoptera* species and served as a preadaptation for their utilization of cactus as host plants. An alternate hypothesis could be that some other trait or traits, physiological perhaps, served as the initial preadaptation for being cactophilic. The possibility that the subgenus *Drosophila* as a whole is characterized by more frequent female remating than the subgenus *Sophophora* is one that remains to be explored. I would like to suggest the hypothesis that the mating systems of the cactophilic *Drosophila* are related to the characteristics of the breeding sites provided by necrotic cactus and suggest certain ways to examine this hypothesis. Rot pockets are usually patchily distributed and each necrosis undergoes a succession of stages. The actual time during which a rot is attractive for oviposition appears to be a highly transitory one (W.B. Heed, pers. comm.). In order for females to take advantage of optimal oviposition substrates, it would appear that they need a ready supply of sperm and eggs. Frequent remating should assure an available sperm supply. But it may also serve to increase egg production (Fig. 2). The possibility that females remate frequently to obtain some component(s) from the ejaculate must be considered. The value of nutrients transferred to females at mating has been nicely documented for the Lepidoptera, and is currently being studied in *Drosophila* in my laboratory (Boggs and Gilbert, 1979; Boggs, 1981).

Another problem apparently confronting these flies is inbreeding. If rots are few and far between or if populations crash in the heat of the summer, populations of related individuals would tend to build up around particular necroses. Laboratory observations suggest that none of the cactophilic species are very tolerant of inbreeding (Cooper, 1964; Jefferson, 1977). One situation that would seem to reduce matings between sibs is the age differential between males and females for reproductive maturation, which is seen in all cactophilic species that have been mated in the laboratory. Further, frequent remating increases the chances of eventually copulating with an unrelated individual. Superimposed upon this might be a mechanism whereby females do not release eggs when inseminated by too close a relative, a sort of "self sterility" situation. The existence of behavioral and physiological mechanisms for preventing inbreeding is also testable and the data from *D. mojavensis* suggests that the other cactophilic species should be studied under inbreeding regimes. The relationship between the mating status of males and females and the timing of migration to a new rot is an unknown which bears on the question of gene flow and the genetic structure of the populations. Furthermore, self sterility mechanisms have been proposed to be of importance in the evolution of mate recognition systems during speciation (Templeton, 1980).

The evolutionary significance of the considerable variability among cactophilic species in the incidence of remating is not clear. Causes may be sought in both mechanistic and evolutionary terms. For example, in certain *repleta* group species from the *mulleri* complex, copulation is followed by the appearance of an insemination reaction, a mass of opaque, non-cellular material that fills the female's uterus and lasts up to 10 hours (Patterson, 1947). The presence of the reaction mass may temporarily prevent remating as well as oviposition in those species which exhibit it. Of the species included in the present study, only *D. mojavensis* and *D. arizonensis* show a reaction mass. Females of these two species do not ordinarily remate for 24 hours. On the other hand, the two *repleta* species which multiply mate in a given morning, *D. nigrospiracula* and *D. hydei*, do not form this mass. Thus the presence or absence of a reaction mass could be one factor underlying the remating propensity of females of certain species. Another factor proposed by Walker (1980) to influence female remating is the size and shape of the female sperm storage organ. Longer storage organs as opposed to spherical ones were suggested to predispose females to more frequent mating. The *repleta* species included in the present study all store their sperm in the ventral receptacle and not in the spermathecae, and all have very long receptacles as compared with other species groups. Females of *D. hydei* have the longest receptacles in the genus (Wasserman, 1960) and together with *D. nigrospiracula* which also has a long receptacle, show the highest incidence of remating ever reported for *Drosophila*. In all of the *repleta* group species examined so far, sperm are found in the

proximal portion of the ventral receptacle immediately after
mating. Within 12 to 24 hours, they are distally located and
the proximal portion is relatively empty. Perhaps the availability of space in the proximal ventral receptacle governs
whether or not a remating will occur. A long ventral receptacle may simply require a number of copulations in order to
fill with sperm. The ventral receptacle does appear to be
innervated and may thus provide one level of control of
female receptivity (Miller, 1950).

Genetic and physiological control of the timing of remating has been most extensively investigated in *D. melanogaster*. Manning (1967) suggested that the act of copulation
itself causes a mechanical block to remating that lasts up to
24 hours in this species. After that time, the inclination
of a female to remate is assumed to depend upon the degree to
which the supply of motile sperm in her ventral receptacle
has been reduced. Reduction in the supply of motile sperm
could result from their being used up during oviposition and/
or from factors in the ejaculate or female reproductive tract
which influence sperm viability (see Gromko *et al.*, 1981 for
an extensive review). Obviously in *D. hydei* and *D. nigrospiracula* no such 24 hour block exists. In all of the cactophilic *Drosophila* species examined in my laboratory, the
utilization of sperm by oviposition appears to have no detectable effect on the timing of remating. Females of any given
cactophilic species appear to remate just as quickly whether
they have been allowed to use up sperm by ovipositing or not.

The role of ecology in the differences in mating systems
between various cactophilic species is more speculative at
this time. One obvious variable is the nature of the breeding sites provided by the preferred host plant of each
species. Saguaro and cardón, the hosts of *nigrospiracula*,
are much larger than organ pipe and agria, which are utilized
primarily by *D. mojavensis*, and senita, utilized by *D. pachea*.
The opuntias provide a different substrate entirely from the
columnar cacti. The size, shape and moisture content of the
cactus undoubtedly influence the features of a rot and the
breeding site it provides. The presence of other species
which might be competing for the same substrate may also
influence mating strategies. *D. pachea* has few problems with
competitors since most species are prevented from using
senita by the toxic alkaloids in this cactus (Heed, 1978).
But *D. nigrospiracula* may have evolved its intense mating
activity as a means of keeping other species from utilizing
its host plant.

In summary, certain mating system features are apparently
characteristic of the cactophilic flies in the *repleta* and

nannoptera groups. Whether or not these were initially important in the ability of those groups to breed on cactus is unknown. Differentiation of mating behavior among the different cactophilic species may in some way reflect ecological pressures confronting each of them. The variability between species has provided models for studying the relationship between mating systems and questions about genetic differentiation between populations, sperm competition, parental investment and sexual selection.

ACKNOWLEDGMENTS

The technical assistance of Ms. Rosa Tang, Mr. Peter Yeager, and Mrs. Michinie Gustafson is gratefully acknowledged. Special thanks are due to Dr. William Heed, University of Arizona for fly strains and stimulating discussions.

REFERENCES

Averhoff, W., and Richardson, R.H. (1974). *Behav. Genet. 4,* 207.
Blum, M.S., and Blum, N.A. (eds.). (1979). "Sexual Selection and Reproductive Competition in Insects." Academic Press, New York.
Boggs, C. (1981). *Evolution 35,* 931.
Boggs, C.L., and Gilbert, L.E. (1979). *Science 206,* 83.
Cooper, J.W. (1964). M.S. Thesis, Univ. of Arizona.
Garcia-Bellido, A. (1964). *Z. Natur. 19,* 491.
Gromko, M., Gilbert, D., and Richmond, R. (1981). *In* "Sperm Competition and the Evaluation of Animal Mating" (R.L. Smith, ed.). Academic Press, New York. (in press).
Heed, W.B. (1978). *In* "Ecology and Genetics: The Interface" (P.F. Brussard, ed.), p. 109. Springer-Verlag, New York.
Jefferson, M.C. (1977). Ph.D. Dissertation, Univ. of Arizona.
Krebs, J.R., and Davies, N.B., (eds.). (1978). "Behavioral Ecology: An Evolutionary Approach." Sinauer, Mass.
Manning, A. (1967). *Anim. Behav. 15,* 234.
Markow, T.A. (1981). *Evolution 35,* 1022.
Merle, J. (1968). *J. Insect Physiol. 14,* 1159.
Miller, A. (1950). *In* "Biology of Drosophila" (M. Demerec, ed.), p. 420. Wiley, New York.
Patterson, J.T. (1947). *Univ. Texas Publ. 4720,* 41.

Patterson, J.T., and Stone, W.S. (1952). "Evolution in the Genus Drosophila." Macmillan, New York.
Pruzan, A. (1976). *Evolution 30*, 130.
Pyle, D.W., and Gromko, M.H. (1978). *Experientia 34*, 449.
Templeton, A.R. (1980). *Genetics 94*, 1011.
Walker, W.E. (1980). *Am. Nat. 115*, 780.
Wasserman, M. (1960). *Proc. natn. Acad. Sci. U.S.A. 46*, 842.
Wasserman, M. (1982). *In* "The Biology and Genetics of Drosophila," Vol. 3b, (M. Ashburner, H.L. Carson, and J.N. Thompson, Jr., eds.). Academic Press, London. (in press).
Wheeler, M.C. (1981). *In* "The Biology and Genetics of Drosophila," Vol. 3a, (M. Ashburner, H.L. Carson and J.N. Thompson, Jr., eds.). Academic Press, London. (in press).

PART V
BIOCHEMISTRY
AND ADAPTATION

19

Tests of the Adaptive Significance of the Alcohol Dehydrogenase Polymorphism in *Drosophila melanogaster*: Paths, Pitfalls and Prospects

John B. Gibson
John G. Oakeshott

Department of Population Biology
Research School of Biological Sciences
The Australian National University
Canberra, A.C.T.

The minimum requirements for evidence sufficient to prove that a particular enzyme polymorphism is adaptive have been set out by Clarke (1975) and Koehn (1978). Phenotypic diversity must exist among genotypes in some measure of molecular function and this functional diversity must be physiologically relevant. Further, a specific locus must be affected by a specific ecological component, *via* the enzyme product of the locus and, critically, the phenotypic differences must be manifest in some measure of fitness.

For the majority of allozyme loci studied in *Drosophila*, the enzymes encoded by the alternate alleles do differ in some of their biochemical properties such as maximum velocity or substrate specificity. But for no allozyme system is there good evidence for the mechanism of selection (Hedrick *et al.*, 1976) and selection coefficients have not been measured in natural populations.

Of all the systems which have been extensively studied the Alcohol dehydrogenase (*Adh*) locus in *D. melanogaster* provides a *cause célèbre* with lessons for studies on enzyme polymorphisms in all species. *D. melanogaster* are nutritionally dependent on yeasts and bacteria (Sang, 1956) and their breeding sites in decaying fruit and vegetable matter are associated with the products of anaerobic respiration (e.g., alcohols). Alcohol dehydrogenase (ADH) is produced in all *D. melanogaster* life stages and at times of peak levels (in third instar larvae and in six to eight day old adult

ECOLOGICAL GENETICS AND EVOLUTION
ISBN 0 12 078820 9

flies) comprises up to 1% of the total soluble protein. The
Adh locus is polymorphic in natural populations and seems,
at first sight, to be most amenable to the type of analyses
outlined by Clarke and Koehn.

Our aim in this paper is to briefly summarise the
relevant evidence on the *Adh* polymorphism in *D. melanogaster*, to point to major ambiguities, and to suggest that
different approaches will be required to solve this problem,
and similar problems in other species such as the cactophilic *Drosophila*.

I   Electrophoretic Variation at the *Adh* Locus in *D.melanogaster*

Five *Adh* alleles encoding electrophoretic ADH variants
have been discovered in natural populations (*US*, *S*, *F*, *F'*,
and *UF*). Additionally, a number of electrophoretically
cryptic (under normal conditions of electrophoresis) thermostability variants have been discovered, including *FCh.D.*
(Gibson *et al.*, 1980), which is probably identical with two
others, *Fr* and *71K*, described by Thörig *et al.* (1975) and
Sampsell (1977). Of these alleles, *F* and *S* are common,
FCh.D. is widespread at relatively low, but polymorphic,
frequencies (Wilks *et al.*, 1980; Gibson, 1982) and *US*, *F'*
and *UF* are probably rare (Chambers, 1981).

The worldwide geographical distribution of the common
Adh alleles is remarkable with the frequency of *S* decreasing
in both hemispheres with increasing distance from the
equator (Vigue and Johnson, 1973; Wilks *et al.*, 1980; Anderson, 1981; Oakeshott *et al.*, 1981). The areal distribution
provides the most persuasive evidence that the *Adh* polymorphism is maintained by some mode of natural selection.

II   Biochemical Differentiation of ADH Variants

The primary amino acid sequences of ADH-S, ADH-FChD,
ADH-F, ADH-F' and ADH-UF are known. Three examples of ADH-S
from geographically separate populations have been
sequenced, and found to be homogeneous. They each differed
from ADH-F by the same single amino acid substitution
(threonine to lysine) (Fletcher *et al.*, 1978; Thatcher,
1980; Chambers *et al.*, 1981a). ADH-FChD differs from ADH-F
by a proline to serine substitution (Chambers *et al.*,

1981a). The nucleotide sequences of F and S are in agreement with the amino acid sequences, although there are two intervening DNA sequences (introns) of 65 and 70 nucleotides within the coding region (Goldberg, 1980; Benyajati *et al.*, 1981).

Turning our attention now to the biochemical properties on which selection might act, it is known that the enzyme catalyses the reversible oxidation of alcohols to aldehydes. Four properties have been found to distinguish F/F and S/S strains (Table I). First, most F/F strains have higher enzyme activity with alcohol substrates, and produce more enzyme molecules at equilibrium, than most S/S strains (Gibson, 1970, 1972; Day *et al.*, 1974; Lewis and Gibson, 1978). In an experiment carried out soon after the discovery of the polymorphism, and testing the hypothesis that heterosis for activity levels might be relevant to the *Adh* polymorphism, Rasmuson *et al.* (1966) found that F/S heterozygotes were intermediate in activity between the two homozygotes. On occasions evidence for some dominance in activity has subsequently been reported (Oakeshott, 1976a).

These differentials in activity levels between F/F and S/S are mirrored, but to a lesser extent, in samples from polymorphic populations in which ADH activity is positively correlated with F frequency (Anderson, 1982). There is also evidence that modifiers, which alter the amount of ADH protein and hence activity, segregate in natural populations (Ward, 1974, 1975; Thompson *et al.*, 1977; Lewis and Gibson, 1978; McDonald and Ayala, 1978; Birley *et al.*, 1980). However, for some modifiers, evidence shows that they probably increase ADH activity only in the presence of an active *Adh* structural gene (Welter and McDonald, 1980).

TABLE I. Properties of Three ADH Variants (data from Chambers et al., 1981b) and Gibson et al., 1980)

Enzyme	*Vmax (units/mg live wt) ethanol substrate at pH8.7*	*Km (mM) at pH8.7*	*Activity ratio (2-propanol /ethanol)*	*Activity (%) after 5 min at 44°C*
ADH-F	86.7	6.5	6.9	3.2
ADH-S	26.6	4.1	4.6	2.4
ADH-FChD	54.6	6.1	4.4	38.9

A second general finding (Table I) is that ADH-F has a higher Km for ethanol than ADH-S (Vigue and Johnson, 1973; Day *et al.*, 1974; Chambers *et al.*, 1981b), although whether the difference remains under *in vivo* conditions is unknown. Third, in purified enzyme preparations, ADH-S has higher thermostability than ADH-F (Vigue and Johnson, 1973) but this difference is not consistent in crude extracts (David, 1977; Wilks *et al.*, 1980). However, in either type of preparation, ADH-FChD is up to ten times more thermostable than the two common forms of the enzyme (Table I) (Gibson *et al.*, 1980). The fourth known phenotypic difference between the variants is in the ratio of activity with 2-propanol to activity with ethanol (Table I). This ratio is higher for ADH-F than for ADH-S and ADH-FChD (Lewis and Gibson, 1978; Gibson *et al.*, 1980; Chambers *et al.*, 1981b).

For each of these four biochemical phenotypes there is evidence for effects of the genetic background (Gibson *et al.*, 1981b) but, in each case, the major part of the effect depends primarily on variation at the *Adh* locus.

III   Physiological Relevance of Phenotypic Diversity

Of the four facets of phenotypic diversity in molecular function, little work has been carried out relating Km or substrate variation to environmental variation. Most emphasis has been placed on ADH activity, although environmental correlates with thermostability have been investigated. There is however considerable doubt about the adaptive significance of the differences in thermostability to ambient temperatures in *Drosophila* habitats. Johnson and Powell (1974) did find that *S/S* homozygotes were at a higher frequency in the 10% of flies which survived a heat shock of ten minutes at 40°C than in the polymorphic test populations before treatment. However the experiments were carried out on standard media, *without* added ethanol, and there was no direct evidence of differences in activity, either before or after heat treatment, between ADH-F and ADH-S.

These experiments have not been adequately repeated, but the discovery of *FCh.D.*, which produces an ADH with much greater thermostability than either ADH-F or ADH-S, helped to resolve the problem. In summary, these data provide no evidence that the frequency of *FCh.D.* is consistently correlated with environmental temperatures, and the differentials in ADH activity are only manifest in extracts subjected to temperatures which are lethal to larvae or adult flies (Wilks *et al.*, 1980).

Experimental evidence for the physiological relevance of the diversity in activity levels between the products of alleles at the *Adh* locus has mostly come from attempts to measure fitness differentials in environments to which extra ethanol has been added, rather than from studies of ethanol tolerance *per se*.

A. *Single Generation Tests*

It has generally been assumed that the main function of ADH in *D. melanogaster* is to detoxify environmental alcohols, although an alternative view that the major role is in lipid metabolism has been championed (Johnson, 1977), but without supporting data. The most direct evidence for the detoxification function comes from studies of *Adh* null mutants lacking ADH activity. Flies homozygous for *Adh* nulls die within a few hours on media made up with 8% ethanol (Grell *et al*., 1968) and are unable to metabolise low (1%), nontoxic, concentrations of ethanol (David, 1977). Conversely, null mutants survive, but flies with ADH activity die, when exposed to 1-penten-3-ol, which is converted by ADH to a toxic ketone (Sofer and Hatkoff, 1972). It has also been shown that, at levels above 3%, ethanol in the media is a metabolic cost to *D. melanogaster* but that up to this concentration the ethanol can be used as an energy resource (Libion-Mannaert *et al*., 1976; Parsons and Stanley, 1981).

In comparisons between naturally occurring *Adh* genotypes it has been shown that the egg to adult survival of *F/F* homozygotes is higher than that of *S/S* homozygotes on media containing more than 6% ethanol (Morgan, 1975; Briscoe *et al*., 1975; Oakeshott, 1976b). However, under the same test conditions, heterozygotes with intermediate levels of ADH activity have been found to show over-dominance in similar measures of fitness (Oakeshott, 1976b). Within the range of activities of *S/S* inbred lines of *D. melanogaster*, there is evidence for a positive correlation between egg to adult survival on ethanol impregnated media (above 6%) and ADH activity (Thompson and Kaiser, 1977). Of particular importance is the evidence (Table II) that the mode of exposure to ethanol is critical for fitness. *S/S* flies have higher survival rates than *F/F* flies after exposure for 45 minutes to ethanol vapour in air, showing that high ADH activity can be disadvantageous under some modes of exposure to ethanol (Oakeshott *et al*., 1980).

TABLE II.   *Survival of Adh Genotypes with Different Exposures of Adults to Ethanol (data from Oakeshott et al., 1980)*

Treatment	% Survival		
	F/F	F/S	S/S
Seven days on 12% ethanol food	63 ± 8	71 ± 5	24 ± 4
One day on 8% ethanol solution	74 ± 6	71 ± 3	42 ± 8
45 minutes in ethanol vapour	29 ± 6	34 ± 6	46 ± 6

A further complication was revealed by Bijlsma-Meeles (1979) who found marked differences *between* experiments in the survival of flies on ethanol medium. Subsequent experiments showed that egg to adult survival of *Adh* genotypes on ethanol media was higher if young eggs were transferred from normal food, due to increased ethanol tolerance and ADH activity in eggs exposed to ethanol. After exposure to ethanol *S/S* eggs have much higher survival, but much lower ADH activity, than unexposed *F/F* eggs.

Additional, but indirect, evidence has come from interspecific comparisons between *Drosophila* species with different levels of ADH. These data (Table III) show a clear positive relationship between levels of ADH and tolerance to ethanol. For example, *D. simulans* (which seems to be monomorphic for an electrophoretic form of ADH) has less activity than *S/S* in *D. melanogaster* and has lower viability on ethanol media (Daggard, 1981).

Apart from showing that ADH probably *can* metabolise environmental ethanol all of these single generation fitness experiments fall short of unequivocally demonstrating the *in vivo* function of ADH. To do this will require labelling experiments but the results will be an important, and necessary, advance in evaluating evidence for the adaptive significance of the *Adh* polymorphism.

TABLE III. *Species Comparisons of ADH Activity, Ethanol Utilization and Tolerance*

	ADH activity	Ethanol utilization /tolerance
Holmes et al. (1980)	D. melanogaster ⌄ 3 fruit-baited Scaptodrosophila spp. ⌄ 3 non-fruit-baited Scaptodrosophila spp.	D. melanogaster ⌄ 3 fruit-baited Scaptodrosophila spp. ⌄ 3 non-fruit-baited Scaptodrosophila spp.
Daggard (1981)	D. melanogaster S/S ⌄ D. simulans ⌄ D. immigrans ⌄ D. busckii	D. melanogaster S/S ⌄ D. simulans ⌄ D. busckii ⌄ D. immigrans

B. *Multiple Generation Experiments*

The clear prediction from the single generation tests (which is certainly not contradicted by the interspecific comparisons) is that F will increase in frequency relative to S in a polymorphic population maintained on ethanol media. It is the results of tests of this prediction that reveal a major paradox in our understanding of the *Adh* polymorphism, for the prediction is not generally borne out. Populations maintained in that way rarely go to fixation for F and indeed do not always show a decrease in S (for references and discussion of examples see Oakeshott and Gibson, 1981).

In a different approach Gibson *et al.* (1979) selected for ethanol tolerance in two populations segregating *Adh* alleles. Neither the frequency of *F* nor the level of ADH activity increased consistently, although the selected flies bred successfully on 12% ethanol media whereas the base populations could not survive on this level of ethanol. However, in tests of single allele *S/S* and *F/F* lines extracted from the same base populations egg to adult viability on ethanol media was positively correlated (r = 0.71, p < 0.01) with the amount of ADH protein assayed by radial immunodiffusion techniques (Lewis and Gibson, 1978).

Van Herrewege and David (1980) reported that selection for ethanol tolerance in two populations segregating *S* and *F* alleles resulted in *F* becoming fixed. In a third population, also selected for ethanol tolerance, *F* was fixed before the selection was imposed. McDonald *et al.* (1977) found that the amount of ADH protein was higher in this selection line than in its base population. But from these experimental results it is not possible to decide whether the increased ADH level caused the increased tolerance, or occurred once tolerance had developed.

Using single female lines maintained on media without added ethanol Wilson *et al.* (1982) were able to measure selection coefficients affecting *Adh* genotypes with values of *S/S*(1.00): *F/S*(1.08): *F/F*(1.08). In later experiments (see data in Table VI below) it was found that the coefficients were similar whether standard or ethanol supplemented medium was used. In both laboratory conditions the measured selection was against the *S* allele although it was not possible to conclude that the selection was directed at variation at the *Adh* locus.

C. *Physiological and Behavioural Components of Ethanol Utilisation*

There are few direct data on the physiological relationships between environmental ethanol, ADH activity and *Adh* gene frequencies. For example, although it has been shown in laboratory experiments that *Drosophila* species can utilise ethanol as an energy source, the utilisation is not solely dependent on ADH activity (Van Herrewege and David, 1980) and catalase activity is implicated (Deltombe-Lietaert *et al.*, 1979). And it has also been shown that alcohol tolerance and utilisation are controlled, at least partly, by different genetic mechanisms (Van Herrewege and David,

1980), for increasing tolerance by selection did not result in a better utilisation of a low concentration of ethanol. McKenzie and Parsons (1974) found that alcohol tolerance segregated to some extent independently of the *Adh* locus in a population of *D. melanogaster* at the Chateau Tahbilk winery.

It is far from clear whether the magnitude of the differences in ADH levels that may distinguish genotypes at the *Adh* locus in natural populations is relevant to the levels of ethanol detected in *D. melanogaster* habitats. Both in the tolerance and utilisation of ethanol there may be physiological thresholds above which the level of ADH is unimportant. This level of ADH may be lower than that produced by *S/S* homozygotes. Also, there are a number of behavioural and physiological strategies that might be used to adapt to ethanol in the environment and which would not depend on ADH activity. To some extent larvae and adult flies can choose their habitats, and adults can select habitats for egg laying to avoid stressful environments. Modification of feeding rates, for which there is evidence for genetic variation (Sewell *et al.*, 1975), and changes in developmental times, are other strategies for avoiding the debilitating effects of ethanol.

D. *Ecological Relevance of Phenotypic Diversity*

Attempts to investigate the *Adh* polymorphism in natural populations have assumed that the differences in activity between *F/F* and *S/S* exist in natural populations and usually have simply compared the allele frequencies in what were assumed to be contrasting environments. The difficulty with this approach is the lack of supporting evidence for the two assumptions. Indeed, it has now been shown that levels of ethanol do not necessarily differ between *Drosophila* habitats "inside" and "outside" wineries (Gibson *et al.*, 1981a). Thus it is not surprising that differences in *F* frequency, between supposedly contrasting environments (but for which data on ethanol levels were not obtained), have not been consistently observed (McKenzie and Parsons, 1972, 1974; Briscoe *et al.*, 1975; Hickey and McLean, 1980; Marks *et al.*, 1980).

McKenzie and McKechnie (1978) did find that ethanol tolerance decreased away from a winery cellar at Chateau Tahbilk but they observed no corresponding changes in *Adh* frequency. Given the survival values that they obtained along the tolerance cline, and the possibility of some dominance, or even overdominance, in ethanol tolerance among *Adh* genotypes (e.g., see Tables II and V), they would not have detected differences in *F* frequency resulting from this selection at the *Adh* locus.

It is also relevant that the frequencies of *D. simulans* are invariably found to be lower "inside" than "outside" wineries, even when there are no differences between the sites in ethanol levels. David (1979) has argued that attractive behaviour towards human constructions goes some way towards explaining the high frequencies of *D. melano-gaster* found in or close to buildings, and this may be relevant to the interpretation of the relative frequencies of *D. simulans* and *D. melanogaster* within wineries.

The north-south clines in *S* frequency correlate with a maximum rainfall variable in both hemispheres (Table IV), but no supporting ecological data are available to explain this consistent, and very marked, relationship (Oakeshott *et al.*, 1981). The *Adh* locus is close to the proximal breakpoint of the cosmopolitan inversion *In(2L)t*, which also shows a north-south cline and with which the *S* allele is in consistent linkage disequilibrium. However the north-south cline in *S* persists in the standard sequence second chromosomes in Australasia (Knibb, personal communication) and in North America (Voelker *et al.*, 1978).

TABLE IV.   *Partial Correlations of S Allele Frequency with Four Climatic Variables. The Maximum and Minimum Temperature and Rainfall Variables are the Average Daily Readings for the Extreme Calendar Months (Data from Oakeshott et al., 1981)*

	Tmax	Tmin	Rmax	Rmin
Australasia	+.31	+.33	+.50[b]	+.38[a]
N. America	+.51[c]	−.18	+.45[c]	+.13
Europe/Asia	−.29	+.30	+.67[c]	+.04

[a] $P < 0.05$
[b] $P < 0.01$
[c] $P < 0.001$

IV   An Attempt to Resolve the Paradox

It is clear that attempts to investigate the adaptive significance of the *Adh* polymorphism have failed because of a lack of information on the physiological role of ADH and on the ecology of *Drosophila*. But it seems to us that even with this information the investigation might not be entirely successful. Our evidence for this argument is drawn from a number of separate observations, made for a variety of other purposes.

For example, one clue came from the results of Gibson *et al.* (1979) referred to above, in which *F* and *S* studied in single allele lines showed fitness differentials on ethanol media but apparently not when segregating in populations exposed to ethanol. Anderson (1982) provided another clue for he showed that ADH activity levels are not consistently related to *F* frequency. He found that differences between *F/F* and *S/S* homozygotes in ADH activity, and in tolerance to ethanol, were significantly reduced in segregating cultures compared to homozygous lines (Table V).

The mechanisms responsible for this apparent genotype x genotype interaction are being investigated, but for our present purposes it is sufficient that they do occur. Anderson (1982) also found that although ADH activity and *F* frequency were correlated in Australasian populations, neither the activity nor the tolerance to ethanol were predictable from data obtained on isolated *F* and *S* lines.

A similar phenomenon was noted by Wills and Nichols (1971, 1972) studying the Octanol dehydrogenase locus in *D. pseudoobscura*. They detected viability differences between genotypes at this locus if octanol was in the culture medium

TABLE V.   *Differential Survival Rates of Adh Genotypes on 12% Ethanol Media (Data from Anderson (1982)*

Lines	% *Survival*		
	F/F	*F/S*	*S/S*
Polymorphic	*37 ± 5*	*38 ± 4*	*25 ± 6*
Fixed	*61 ± 7*	-	*14 ± 9*

and when inbred populations were tested, but not when the genetic background for the *Odh* locus was heterozygous.

Re-examination of the experimental protocols used for the published multi-generation experiments on the *Adh* polymorphism reveal a relationship between the fate of *S* on ethanol media and whether the base populations were inbred laboratory stocks or mass collections recently derived from wild populations. Most of the experiments in which *F* increased on ethanol media had been initiated from relatively inbred stocks. And using the general plan of the experiments of Wilson *et al.* (1982), we have recently obtained consonant results in comparisons beween populations derived from single female lines and those from mass cultures; the frequency of F increased consistently only in the single female lines, regardless of whether or not ethanol was added to the media (Table VI).

The conclusion that emerges from the juxtaposition of these separate pieces of information is that any role of the *Adh* locus in ethanol tolerance depends, not only on the effects of modifiers, but also to a significant extent on the genetic architecture of the population investigated. This will be recognised as merely a new example of an old problem. Lerner (1954), Mather (1955), Lewontin (1956) and Parsons (1959) all provided experimental evidence

TABLE VI. *Relative Fitnesses of Adh Genotypes in Different Experimental Conditions*

Populations	% Ethanol in food	Relative fitness coefficients		
		F/F	F/S	S/S
54 isofemale lines	0%	1.08	1.08	1.00
4 isofemale lines	0%, 3%, 9%	1.09	1.04	1.00
7 mass collections	0%, 3%, 9%	1.00	1.00	1.00

demonstrating positive relationships between genetic homeostasis in outbred populations and resistance to environmental fluctuations. Specifically, inbred lines lose this
capacity and show considerable phenotypic variation between
environments.

Other than the experiments of Wills and Nichols (1971,
1972) the topic has received little attention in relation to
specific allozyme loci, although Beardmore and Shami (1979)
have found evidence that optimum and non-optimum phenotypes
for a trait under stabilising selection differ in average
genotypes at four allozyme loci in *Poecilia reticulata*. And
in experiments with *D. buzzatii*, Gunawan (1981) found some
evidence for a positive relationship between measured levels
of heterozygosity and the heritability of six bristle
characters. Nevertheless these observations *in toto* point
to a general problem that will arise in studies of any
enzyme polymorphism. Comparison of isolated homozygous
lines (as in Cavener, 1979) is clearly inappropriate, but in
studies of viability differences in segregating material
investigators usually try to minimise the effects of the
genetic background so that effects ascribable mainly to the
allozyme locus can be observed. It seems likely that this
technique, in which the locus investigated segregates in a
relatively inbred background, provides data on fitnesses
which may be irrelevant to natural populations.

This effect may arise from a breakdown of genetic
homeostasis, but the cause could be a reduction in overall
heterozygosity or the disruption of a genetic architecture
maintained in the wild by stabilising selection. Whatever
the cause, it does seem reasonable to argue that effects at
single loci that can only be identified in inbred lines will
be of limited biological importance in natural populations.
In the case of the *Adh* polymorphism there is good evidence
for this point of view as comparisons of populations in
environments which do differ in ethanol levels show that the
populations do not differ in *Adh* frequencies (Gibson *et al.*,
1981a).

A major problem for the next phase of research will be
to devise experimental approaches which will allow for the
detection of selective effects at individual allozyme loci
without so changing their genetic backgrounds that the
effects observed are uncharacteristic of natural populations. It clearly will be necessary to put much greater
emphasis on '*in vivo*' ecological studies of populations in
nature.

REFERENCES

Anderson, P.R. (1981). *In* "Genetic Studies of Drosophila Populations" (J.B. Gibson and J.G. Oakeshott, eds.), p. 237. Australian National University, Canberra.

Anderson, D.G. (1982). *In* "Proceedings of the VII European Drosophila Research Conference". Plenum, New York (in press).

Beardmore, J.A., and Shami, S.A. (1979). *Aquila. Ser. Zool. 20*, 100.

Benyajati, C., Place, A.R., Powers, D.A., and Sofer, W. (1981). *Proc. natn. Acad. Sci. U.S.A. 78*, 2717.

Bijlsma-Meeles, E. (1979). *Heredity 42*, 79.

Birley, A.J., Marson, A., and Phillips, L.C. (1980). *Heredity 44*, 251.

Briscoe, D.A., Robertson, A., and Malpica, J. (1975). *Nature, Lond. 255*, 248.

Cavener, D. (1979). *Behav. Genet. 9*, 359.

Chambers, G.K. (1981). *In* "Genetic Studies of Drosophila Populations" (J.B. Gibson and J.G. Oakeshott, eds.), p. 77. Australian National University, Canberra.

Chambers, G.K., Laver, W.G., Campbell, S., and Gibson, J.B. (1981a). *Proc. natn. Acad. Sci. U.S.A. 78*, 103.

Chambers, G.K., Wilks, A.V., and Gibson, J.B. (1981b). *Aust. J. Biol. Sci. 34*, 625.

Clarke, B. (1975). *Genetics 79*, 101.

Daggard, G.E. (1981). *In* "Genetic Studies of Drosophila Populations" (J.B. Gibson and J.G. Oakeshott, eds.), p. 59. Australian National University, Canberra.

David, J. (1977). *Ann. Biol. 16*, 451.

David, J. (1979). *Experientia 35*, 1436.

Day, T.H., Hillier, P.C., and Clarke, B. (1974). *Biochem. Genet. 11*, 141.

Deltombe-Lietaert, M.C., Delcour, J., Lenelle-Monfort, N., and Elens, A. (1979). *Experientia 35*, 579.

Fletcher, T.S., Ayala, F.J., Thatcher, D.R., and Chambers, G.K. (1978). *Proc. natn. Acad. Sci. U.S.A. 75*, 5609.

Gibson, J.B. (1970). *Nature, Lond. 227*, 959.

Gibson, J.B. (1972). *Experientia 28*, 975.

Gibson, J.B. (1982). *In* "Proceedings VII European Drosophila Research Conference". Plenum, New York (in press).

Gibson, J.B., Lewis, N., Adena, M.A., and Wilson, S.R. (1979). *Aust. J. Biol. Sci. 32*, 387.

Gibson, J.B., Chambers, G.K., Wilks, A.V., and Oakeshott, J.G. (1980). *Aust. J. Biol. Sci. 33*, 479.

Gibson, J.B., May, T.W., and Wilks, A.V. (1981a). *Oecologia*
    51, 191.
Gibson, J.B., Wilks, A.V., and Chambers, G.K. (1981b). *In*
    "Genetic Studies of Drosophila Populations" (J.B. Gibson
    and J.G. Oakeshott, eds.), p. 251. Australian National
    University, Canberra.
Goldberg, D.A. (1980). *Proc. natn. Acad. Sci. U.S.A. 77*,
    5794.
Grell, E.H., Jacobson, K.B., and Murphy, J.B. (1968). *Ann.*
    N.Y. Acad. Sci. 151, 441.
Gunawan, B. (1981). *In* "Genetic Studies of Drosophila
    Populations" (J.B. Gibson and J.G. Oakeshott, eds.),
    p. 147. Australian National University, Canberra.
Hedrick, P.W., Ginevan, M.R., and Ewing, E.P. (1976). *Annu.*
    Rev. Ecol. & Syst. 7, 1.
Hickey, D.A., and McLean, M.D. (1980). *Genet. Res. 36*, 11.
Holmes, R.S., Moxon, L.N., and Parsons, P.A. (1980). *J.*
    exp. Zool. 214, 199.
Johnson, F.M., and Powell, A. (1974). *Proc. natn. Acad.*
    Sci. U.S.A. 71, 1783.
Johnson, G. (1977). *Annu. Rev. Ecol. & Syst. 8*, 309.
Koehn, R.K. (1978). *In* "Ecological Genetics: The Interface"
    (P.F. Brussard, ed.), p. 51. Springer, Berlin-
    Heidelberg-New York.
Lerner, I. (1954). "Genetic Homeostasis." Oliver and Boyd,
    Edinburgh.
Lewis, N., and Gibson, J.B. (1978). *Biochem. Genet. 16*,
    159.
Lewontin, R.C. (1956). *Am. Nat. 90*, 237.
Libion-Mannaert, M., Delcour, J., Deltombe-Leitaert, N.,
    Lenelle-Monfort, N., and Elens, A. (1976). *Experientia*
    32, 22.
Marks, R.W., Brittnacher, J.G., McDonald, J.F., Prout, T.,
    and Ayala, F.J. (1980). *Oecologia 47*, 141.
Mather, K. (1955). *Cold Spring Harbour Symp. Quant. Biol.*
    20, 158.
McDonald, J.F., and Ayala, F.J. (1978). *Genetics 89*, 371.
McDonald, J.F., Chambers, G.K., David, J., and Ayala, F.J.
    (1977). *Proc. natn. Acad. Sci. U.S.A. 74*, 4562.
McKenzie, J.A., and McKechnie, S.W. (1978). *Nature, Lond.*
    272, 75.
McKenzie, J.A., and Parsons, P.A. (1972). *Oecologia 10*,
    373.
McKenzie, J.A., and Parsons, P.A. (1974). *Genetics 77*,
    385.
Morgan, P. (1975). *Heredity 34*, 124.

Oakeshott J.G. (1976a). *Aust. J. Biol. Sci. 29*, 365.

Oakeshott, J.G. (1976b). *Genet. Res. 26*, 265.

Oakeshott, J.G., and Gibson, J.B. (1981). *In* "Genetic Studies of Drosophila Populations" (J.B. Gibson and J.G. Oakeshott, eds.), p. 103. Australian National University, Canberra.

Oakeshott, J.G., Gibson, J.B., Anderson, P.R., and Champ, A. (1980). *Aust. J. Biol. Sci. 33*, 105.

Oakeshott, J.G., Gibson, J.B., Anderson, P.R., Knibb, W.R., Anderson, D.G., and Chambers, G.K. (1981). *Evolution* (in press).

Parsons P.A. (1959). *Genetics 44*, 1325.

Parsons, P.A., and Stanley, S.M. (1981). *In* "Genetic Studies of Drosophila Populations" (J.B. Gibson and J.G. Oakeshott, eds.), p. 47. Australian National Univeristy, Canberra.

Rasmuson, B., Nilson, L.R., Rasmuson, M., and Zeppezauer, E. (1966). *Hereditas 56*, 313.

Sampsell, B. (1977). *Biochem. Genet. 15*, 971.

Sang, J.H. (1956). *J. Exp. Biol. 33*, 45.

Sewell, D., Burnett, B., and Connolly, K. (1975). *Genet. Res. 24*, 163.

Sofer, W., and Hatkoff, M.A. (1972). *Genetics 72*, 545.

Thatcher, D.R. (1980). *Biochem. J. 187*, 875.

Thompson, J.N., and Kaiser, T.N. (1977). *Heredity 38*, 191.

Thompson, J.N., Ashburner, M., and Woodruff, R.C. (1977). *Nature, Lond. 270*, 363.

Thörig, G.E.W., Schoone, A.A., and Scharloo, W. (1975). *Biochem. Genet. 13*, 721.

Van Herrewege, J., and David, J.R. (1980). *Heredity 44*, 229.

Vigue, C.L., and Johnson, F.M. (1973). *Biochem. Genet. 9*, 213.

Voelker, R.A., Cockerham, C.C., Johnson, F.M., Schaffer, H.E., Mukai, T., and Mettler, L.E. (1978). *Genetics 88*, 515.

Ward, R.D. (1974). *Biochem. Genet. 12*, 449.

Ward, R.D. (1975). *Genet. Res. 26*, 81.

Welter, J.R., and McDonald, J.F. (1980). *Drosophila Inf. Serv. 55*, 143.

Wilks, A.V., Gibson, J.B., Oakeshott, J.G., and Chambers, G.K. (1980). *Aust. J. Biol. Sci. 33*, 575.

Wills, C., and Nichols, L. (1971). *Nature, Lond. 233*, 123.

Wills, C., and Nichols, L. (1972). *Proc. natn. Acad. Sci. U.S.A. 69*, 323.

Wilson, S.R., Oakeshott, J.G., Gibson, J.B., and Anderson, P.R. (1982). *Genetics* (in press).

20
Biochemical Genetics of the Alcohol Longevity Response of *Drosophila mojavensis*[1]

Philip Batterham
William T. Starmer
David T. Sullivan

Department of Biology
Syracuse University
Syracuse, New York

The genes encoding the enzymes which control alcohol metabolism in *Drosophila* have been extensively studied using a variety of approaches. *D. mojavensis* is a particularly attractive species to which these approaches should be adapted. In this species, extensive ecological analysis indicates that at least one enzyme involved in alcohol metabolism, alcohol dehydrogenase (ADH), is responding directly to environmental parameters. *Adh* gene frequencies in natural populations are correlated with levels of alcohols in host plants, and flies of the various *Adh* genotypes display differential capabilities in extending their longevities in ethanol atmospheres (Starmer *et al.*, 1977; Heed, 1978). These data indicate the possibility of establishing a causal link between environmental alcohols and defined genetic, biochemical, physiological and behavioural responses, thus providing a comprehensive understanding of how these flies utilize alcohols to extend their longevity. With this as our ultimate objective we are studying the role of the ADH enzyme in the observed alcohol-dependent longevity extension. The studies we present here more fully describe the longevity extension response, its relationship to *Adh* genotype, and our initial observations aimed at a complete biochemical, developmental and genetic characterization of the *Adh* locus and its product (ADH) in *D. mojavensis*.

[1]*This study was supported by USPHS grant AG-02064*

307

I   Adult Longevity Analysis

Longevity analyses similar in design to those of Starmer *et al.* (1977) were conducted on a series of populations, using a set of volatile compounds, to provide a more complete picture of the physiological capabilities of *D. mojavensis*. The populations examined were collected at Vallecito (California, U.S.A.), Santa Rosa mountains (Arizona, U.S.A.) and at various sites in Baja California and Sonora (Mexico). Four to seven day old adult females were tested for their longevity response to various atmospheric concentrations of ethanol, 2-propanol, acetic acid, acetone, and to the water control. The results of these experiments are shown in Figures 1-4. Most strains are able to increase their longevity on low concentrations of each of the four volatiles. The response to ethanol is in agreement with the initial observations of Starmer *et al.* (1977). Ethanol increases longevity at concentrations of up to 6%, although optimal longevity is achieved at concentrations between 2% and 4% for most strains (Figure 1). A similar response is observed with acetic acid (Figure 2). These results would be expected if *D. mojavensis* converts ethanol into acetaldehyde and then into acetic acid. Figure 3 shows that some strains increase their longevity on 2% 2-propanol, while other strains do not. Furthermore, some strains increase their longevity when exposed to acetone concentrations below 2% (Figure 4). Those strains which were collected in Baja California, where the insect utilizes agria cactus as a host plant, were able to extend their longevity with either 2-propanol or acetone. Since agria cactus rots are known to have substantial concentrations of these compounds in them (Vacek, 1979), and the Baja strains of *D. mojavensis* tested have a high Adh^F frequency (p = 0.99), it is possible that the observed longevity increases in these strains are due to the superior ability of the ADH-FAST allozyme in utilizing 2-propanol as a substrate. However, this hypothesis does not explain the observed longevity increases with acetone, since acetone is not known to be a substrate of ADH or any other *Drosophila* enzyme. In fact, acetone inhibits ADH activity in *D. melanogaster* (Papel *et al.*, 1979) and in *D. mojavensis* (Batterham and Sullivan, unpublished). We therefore consider it likely that acetone utilization is mediated by yeasts which are resident within the fly.

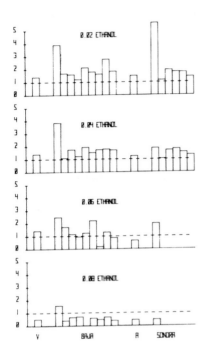

FIGURE 1. Longevity
response of various geographic strains of D. mojavensis
to atmospheric ethanol. Longevity response is expressed as
the ratio of the mean longevity of 20 females tested on
ethanol to the mean longevity
of 20 females tested on the
water control. Ratios in excess of 1.10 indicate that the
longevity is significantly increased, and ratios less than
0.90 indicate that longevity
is significantly decreased in
comparison to the water control. 95% confidence limits
were calculated using the
method of Bliss (1967). The
strains tested were from Vallecito, Southern California
(V), Santa Rosa, Arizona (A),
and numerous locations in Baja
California and Sonora in
Mexico.

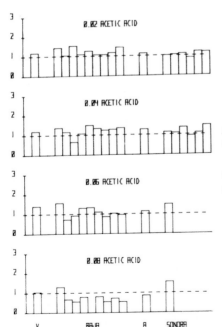

FIGURE 2. Longevity
response of geographic strains
to atmospheric acetic acid.
(Experimental methods and presentation of data are the same
as for Figure 1).

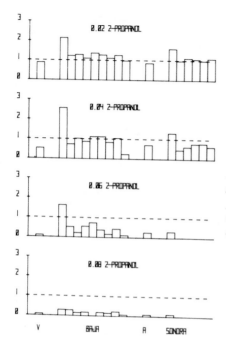

FIGURE 3. *Longevity response of geographic strains to atmospheric 2-propanol. (Experimental methods and presentation of data are the same as for Figure 1).*

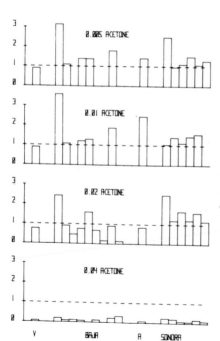

FIGURE 4. *Longevity response of geographic strains to atmospheric acetone. (Experimental methods and presentation of data are the same as for Figure 1).*

II   The Effect of Yeast in Determining Adult Longevity Response

The observed longevity responses in *D. mojavensis* may be
due to the combined contributions of the metabolic systems
of the fly and of the yeasts which are resident in the crop
and gut of the fly. As an initial step in partitioning the
relative contributions of these systems, longevity responses
were compared between axenic and non-axenic strains (Figure
5). Considerable variation was observed in results obtained
between the populations tested. The presence or absence of
yeasts did not significantly affect the longevity responses
of the Vallecito (California) and Catavina (Baja) strains,
whereas significant microorganism-dependent effects were
observed for the San Telmo (Baja) population. These

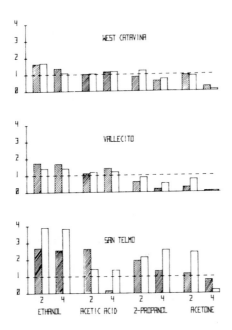

*FIGURE 5. Comparison of longevity extensions of three
geographic strains under axenic culture (open) or with resident yeast flora (hatched). Experimental methods and
presentation of data are as for Figure 1.*

differences could indicate variation in the yeast flora present in each of the three non-axenic strains. Furthermore,
there were differences in the ability of the three axenic
strains to extend longevity using some compounds, indicating
differences in the flies themselves. In spite of this variation some general trends were observed. Axenic strains
from each of the three populations showed a significant increase in longevity on ethanol when compared with the water
control. This is in agreement with the observations of
Starmer *et al.* (1977) which suggested that *D. mojavensis*
is able to metabolize ethanol yielding energy for extended
longevity. The San Telmo strain also displayed a yeast independent ability to extend longevity on 2% 2-propanol and
2% acetic acid. None of the axenic strains were able to
increase their longevity on acetone, which suggests that previously described longevity increases may have been yeast
mediated.

III   The Effect of *Adh* Genotype on Longevity Extension in Ethanol Atmospheres

Adh Fast and Slow homozygotes of 3 different ages (1,
7 and 14 days old) were tested at 5 different atmospheric
ethanol concentrations (0, 2, 4, 6 and 8 percent). The data
are presented in Figure 6 and an analysis of variance is
shown in Table I. The analysis of variance indicates some
statistical interactions are significant, but the majority
of the significant effects are due either to concentration
of ethanol, genotype or age. Figure 6 depicts the genotype
by concentration interaction and shows the genotypes have
essentially the same longevity for water controls. The significant feature of this experiment is that the Adh^S/Adh^S
genotypes exhibit greater longevities in the range of ethanol
concentrations known to be in the environment. This observation is consistent with earlier observations of different
longevities for the different *Adh* genotypes when challenged
with ethanol, and is supportive of the hypothesis that the
ADH-SLOW allozyme is better able to utilize ethanol as a

TABLE I. *Analysis of Variance for Genotype (Adh^F vs Adh^S), Age (1 Day, 7 Days, and 14 Days Old) and Concentration of Ethanol in the Atmosphere (0,2,4,6 and 8 Percent). Observations are Vial Longevities in Hours (5 Females per Vial). Four Vials were Observed for Each Genotype, Age Concentration Combination*

Sources	*d.f.*	*S.S.*	*M.S.*	*F*	*P*
Total	*119*				
Genotype	*1*	*15,229*	*15,229*	*42.5*	*< 0.001*
Age	*2*	*15,679*	*7,839*	*21.9*	*< 0.001*
Concentration	*4*	*188,665*	*47,166*	*131.6*	*< 0.001*
Genotype x Age	*2*	*1,309*	*655*	*1.8*	*0.170*
Genotype x Conc.	*4*	*4,647*	*1,162*	*3.2*	*0.020*
Age x Conc.	*8*	*18,416*	*2,427*	*6.8*	*< 0.001*
Geno x Age x Conc.	*8*	*5,767*	*721*	*2.0*	*0.060*
Error	*90*	*32,249*	*358*		

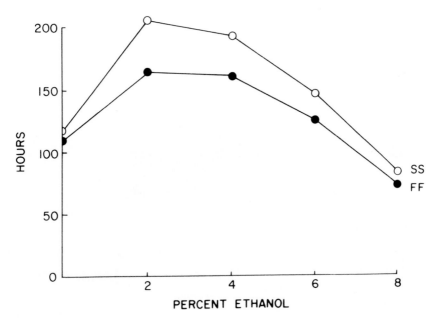

FIGURE 6. *Mean longevity of strains homozygous for the FAST (FF) and SLOW (SS) alleles at the Adh locus at various ethanol concentrations. The two strains used were isofemale lines derived from the Vallecito population. The least significant difference for this graph is approximately 16 hrs.*

substrate than the ADH-FAST allozyme (Starmer *et al.*, 1977; Heed, 1978). However, a biochemical analysis of the respective allozymes will be required to directly implicate the ADH enzyme in this effect.

IV   Evidence for Duplicate *Adh* Loci in *D. mojavensis*

During our initial electrophoretic analysis of the alcohol metabolizing enzymes of *D. mojavensis*, we occasionally noticed what appeared to be an additional band of ADH activity which has lower anodal mobility. Following the observations of Barker and Mulley (1976) in *D. buzzatii*, we hypothesized the existence of duplicate genes for ADH in *D. mojavensis*. Evidence supporting this hypothesis is presented in Figure 7. In adults, the extra band of ADH activity is apparent only in females. We have adopted the

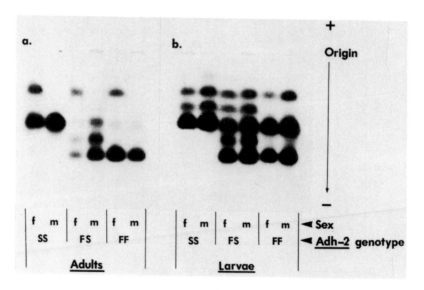

FIGURE 7.   *Agarose gel electropherograms stained for ADH activity using 2-Propanol and NAD+ as substrates.*

convention of naming this female specific band ADH-1, and
the other band(s), ADH-2. The electrophoretic mobility or
presence of ADH-1 is not affected by allelism at the *Adh-2*
locus.

The electrophoretic pattern of ADH bands in larvae is
more complex than that of adults. In larvae of both sexes
three major bands of activity are found in *Adh-2* homozygotes.
One of these is ADH-2 which exists in either the FAST or
SLOW allozymic forms. A second band has the mobility of the
ADH-1 isozyme of adult females. The third band is found
between ADH-1 and ADH-2, and its mobility is affected by the
mobility of the respective ADH-2 allozymes. This band of
activity probably indicates the presence of a heterodimer
composed of one subunit product from each of two loci. The
protein subunit composition of this putative heterodimer is
currently being studied with purified isozymes.

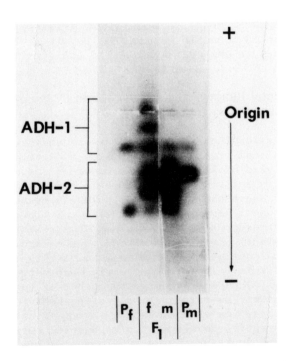

FIGURE 8. *ADH gel staining patterns of the D. mojavensis*
parental female (P$_f$), the D. arizonensis parental male (P$_m$),
and male (m) and female (f) F$_1$ progeny.

Further evidence for the existence of duplicate *Adh* loci has been obtained from the analysis of progeny obtained by crossing *D. mojavensis* females with *D. arizonensis* males (Figure 8). These sibling species differ in the electrophoretic mobility of both the ADH-1 and ADH-2 isozymes. F_1 female progeny show a six banded pattern which appears to be composed of a three banded heterozygous pattern for each of the two loci. As expected F_1 males show only the three banded ADH-2 heterozygous pattern.

From these results and others we have concluded that *D. mojavensis* has duplicate genes each of which code for an ADH subunit that can associate to form an active ADH dimer. Furthermore the products of the separate loci are able to associate to form a heterodimer. This is only possible when the isozymic subunits are synthesized in the same cell. Therefore, the presence of the ADH-1/ADH-2 heterodimer can be used to ascertain whether both genes coordinately function in the same cell. Re-examination of the ADH pattern found in adult females suggests that coordinate function of the *Adh-1* and *Adh-2* genes does not occur (Figure 7). Since no heterodimer is found in mature females, it is likely that *Adh-1* functions in one cell type(s) and *Adh-2* in others.

V   Developmental Biology of *Adh* Function in *D. mojavensis*

The suggestion that the duplicate *Adh* genes are differentially regulated prompted us to pursue a detailed developmental analysis of ADH activity. Figure 9 shows a profile of the specific activity of total ADH in the course of development. High and steadily increasing levels of activity are present throughout larval and pupal life. Following emergence activity drops to lower values. This relatively higher larval enzyme activity may be related to the larvae being intimately and continuously exposed to a variety of alcohols in the cactus rots. The striking difference between males and females in ADH specific activity is not as yet fully explicable but may be related to differing behavioural or ecological strategies of the sexes.

The differential development of ADH-1 and ADH-2 activity during larval life is presented in Figure 10. It is apparent that ADH-1 is present at an earlier point in development than ADH-2 (i.e., eggs and first instar larvae). Whether the early presence of ADH-1 is due to *de novo* synthesis or carryover from eggs is not yet clear. Experiments designed to distinguish between these alternatives are in progress.

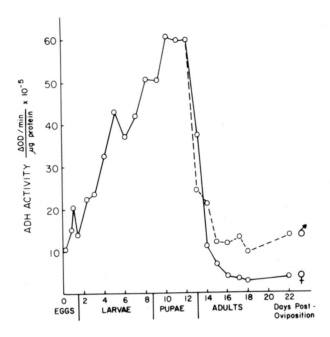

FIGURE 9. *Specific activity of ADH during development in a Vallecito isogenic Adh-2^F strain.*

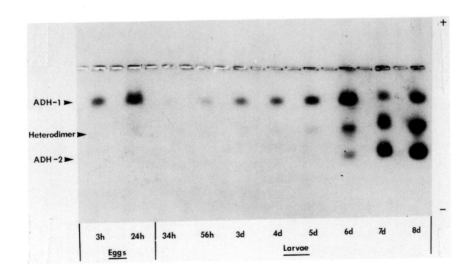

FIGURE 10. *Expression of ADH-1 and ADH-2 activity during larval development.*

The expression of *Adh-1* and *Adh-2* in pupae and adults of each sex is shown in Figures 11 and 12. Figure 11 shows the isozyme content of males. In pupae and newly eclosed adults ADH-1, ADH-2 and the heterodimer are present; in mature males only ADH-2 is present. We have determined by dissection experiments that the ADH-1 and heterodimer of immature adult males are primarily localized in residual larval fat body, and that the loss of these isozymes coincides with the loss of this tissue. In females the pupal pattern is similar to males (Figure 12). However, following emergence the heterodimer disappears, yet ADH-1 persists and even increases in intensity in mature females 9 days post eclosion. The loss of heterodimer in young females is due to the loss of larval fat body. However, ADH-1 expression occurs in cells that do not detectably express ADH-2. This is demonstrated by dissection experiments which show ADH-1 in dissected eggs, but only ADH-2 in the remaining female carcass (Figure 13).

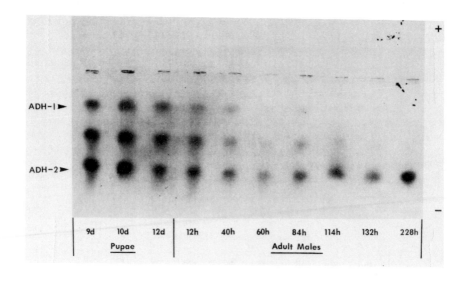

FIGURE 11. *Expression of ADH-1 and ADH-2 activity in pupae and adult males.*

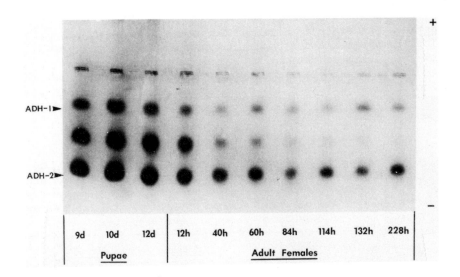

FIGURE 12.   *Expression of ADH-1 and ADH-2 activity in pupae and adult females.*

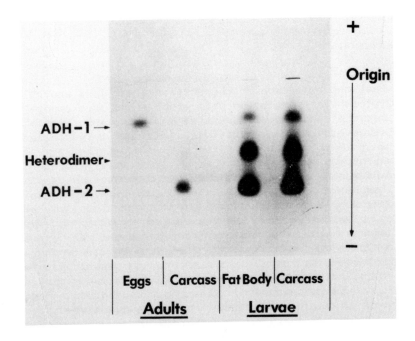

FIGURE 13.   *ADH isozyme content of dissected tissues.*

The expression of the two *Adh* genes in third instar larvae is summarized in Table II. Striking is the observation that most tissues appear to have equal ratios of ADH-1 and ADH-2, but in the gut ADH-1 greatly predominates.

In adults ADH-2 is found in the head, thorax and abdomen. Of particular interest is the observation that ADH-2 alone is found in male reproductive tissue, while only ADH-1 is found in female reproductive tissue.

TABLE II.   *Distribution of ADH-1 and ADH-2 in Eight Day Larvae as Determined by Tissue Dissection and Electrophoresis. Isozyme Staining Intensities on Gels were Scored in Arbitrary Units (+) as a Measure of Isozyme Levels*

	ADH-1	ADH-2
Gut	+ + +	+
Brain	-	-
Tubules	+	+
Salivary gland	+	+
Fat body	+ + + +	+ + + +
Cuticle	+ +	+ +

VI   Conclusion

We feel the existence of the duplicate *Adh* loci and their differential regulation will provide a unique opportunity for the study of differential gene function at the molecular level. Several questions are immediately apparent. First, what is the genetic relationship between the loci? One might expect that the duplication is of fairly recent origin and the genes may be closely linked. Oakeshott *et al.* (1982) demonstrated tight linkage for duplicated *Adh* genes in another *mulleri* sub-group species, *D. buzzatii*. Some preliminary evidence derived from the cross between *D. mojavensis* and *D. arizonensis* indicates that the duplicated *Adh* genes are also linked in *D. mojavensis*. Recently an electrophoretic variant of ADH-1 has been identified and the genetic mapping of *Adh-1* and *Adh-2* with respect to one another is in progress. Secondly, it is reasonable to expect that cloned sequences of *D. mojavensis Adh* - DNA might be isolated and analyzed to examine the evolutionary relationship of these genes with respect to structural and regulatory sequences.

A fundamental problem that must be pursued is to determine the ecological relevance of the *Adh* duplication. A noteworthy point is that larvae express both genes in the same cells and therefore become permanent functional heterozygotes in that both homodimers and the heterodimer are present. If this is ecologically important it might be evidenced in the catalytic properties of the respective enzyme molecules. We have begun studies aimed at investigating this question. We are able to separate and purify each of the homodimers and the heterodimer. These are being subjected to a kinetic and structural analysis.

ACKNOWLEDGMENTS

We are grateful to Peter Brown, Elliott Gritz and Rory Houghtalen for technical assistance at various stages of the work, and to Dr. Don Gilbert for assistance with statistical analysis.

REFERENCES

Barker, J.S.F., and Mulley, J.C. (1976). *Evolution 30*, 213.
Bliss, C.I. (1967). "Statistics in Biology", Vol. 1, p. 219. McGraw-Hill, New York.
Heed, W.B. (1978). *In* "Ecological Genetics: The Interface" (P.F. Brussard, ed.), p. 109. Springer-Verlag, New York.
Oakeshott, J.G., Chambers, G.K., East, P.D., Gibson, J.B., and Barker, J.S.F. (1982). *Aust. J. Biol. Sci. 35*, 73.
Papel, I., Henderson, M., Van Herrewege, J., David, J., and Sofer, W. (1979). *Biochem. Genet. 17*, 553.
Starmer, W.T., Heed, W.B., and Rockwood-Sluss, E.S. (1977). *Proc. natn. Acad. Sci. U.S.A. 74*, 387.
Vacek, D.C. (1979). Ph.D. Dissertation, Univ. of Arizona.

21
Non-Specific Esterases of *Drosophila buzzatii*[1]

P. D. East

Department of Animal Science
University of New England
Armidale, New South Wales

I Introduction

Non-specific esterases are ubiquitous in insects, and
have been the subject of considerable study for many years.
A variety of electrophoretic techniques has been used for the
detection of these enzymes, and most species possess a multi-
plicity of esterase activities. Genetic variation at loci
coding for esterases is also common (Nevo, 1978).

In common with many other *Drosophila* species (e.g.,
Johnson *et al.*, 1966; Sasaki and Narise, 1978), *D. buzzatii*
possesses two predominant esterase activities, one which preferentially hydrolyses 1-naphthylacetate (*Esterase-2*) and the
other, 2-naphthylacetate (*Esterase-1*). Polyacrylamide (PAG)
disc gel electrophoresis commonly reveals 15 to 20 zones of
esterase activity in *D. buzzatii* adults. However, densitometric scanning of these gels indicates that 50-70% of this
activity can be accounted for by only two zones, *Esterase-1*
and *Esterase-2*. Apart from their biochemical interest, these
two enzymes also are of genetic interest because we have considerable circumstantial evidence, especially for the
Esterase-2 locus, that the polymorphisms at these loci are
not neutral with respect to selection. Both loci have shown
associations with a number of macroenvironmental variables in
a multivariate analysis of spatial variation (Mulley *et al.*,

[1]*Supported by a grant from the Australian Research Grants
Scheme to J.S.F. Barker.*

ECOLOGICAL GENETICS AND EVOLUTION
ISBN 0 12 078820 9

1979), and microenvironmental variables in analyses of rot emergences (Barker, 1982). Selection has been implicated in the maintenance of *Esterase-2* polymorphism in a gene frequency perturbation study in a natural population (Barker and East, 1980) and this locus has also shown non-random temporal variation in one population (Barker, 1981), and significant association with yeasts in a field study (Barker *et al.*, 1981). Finally, genotypes at both esterase loci showed significant differences as a result of temperature shocks, *Esterase-1* showing effects under cold shock, and *Esterase-2* under heat shock (Watt, 1981).

As a part of our attempt to determine whether selection actually is acting on these loci, and if so, to determine the mode of action of that selection, we have undertaken a detailed study of the biochemical and physiological characteristics of the allozymes produced by these loci.

II   Ontogeny, Tissue and Sub-Cellular Distribution

As the function of the esterases is unknown, analysis of their temporal and spatial distribution within the organism seemed desirable. The sub-cellular distribution was examined by differential centrifugation of a crude homogenate and although each fraction produced a characteristic spectrum of activities, *Esterase-1* and *Esterase-2* appear to be predominantly cytoplasmic or extracellular in origin.

The tissue distribution was examined in late third instar larvae. The relative contributions of the various tissues were assessed by assaying total esterase and protein, then partitioning activities to particular enzymes by PAG electrophoresis and densitometric scanning of the stained gels (Table I). There are several points to be made; firstly, in late third instar larvae there are three predominant esterases rather than two as in the adult. Secondly, *Esterase-1* is essentially unique to the haemolymph, and *Esterase-2*, though widespread, is predominantly (65-70%) located in the alimentary tract, with high specific activity in the midgut and malpighian tubules. The majority of the remaining *Esterase-2* is located in the fat body. This may be due to some endogenous function of the enzyme in this tissue, or it may be that the enzyme is synthesised there. Finally, the extra esterase, designated *Esterase-J*, since it only occurs in juvenile forms, is located almost exclusively in the carcass.

TABLE I. *Tissue Distribution of Non-specific Esterases*[a]

Tissue type	Specific activity	Total activity	% total activity	Est-2 activity	Est-2 as % total Est-2 recovered activity	Est-1 activity	Est-1 as % total Est-1 recovered activity	Est-J activity	Est-J as % total Est-J recovered activity
Haemolymph	0.03	1.00	–[b]	–	0	0.86	100	–	–
Brain	0.83	3.42	4.4	2.02	3.1	–	–	1.4	11.9
Salivary gland	0.86	2.36	3.1	2.36	3.6	–	–	–	–
Fat body	0.54	15.36	19.9	15.00	23.1	–	–	–	–
Foregut	1.96	17.33	22.5	17.09	26.4	–	–	–	–
Midgut	2.29	22.22	28.8	22.22	34.3	–	–	–	–
Hindgut	1.65	0.57	0.7	0.57	0.9	–	–	–	–
Malpighian tubules	3.08	3.82	5.0	3.82	5.9	–	–	–	–
Carcass	0.41	12.09	15.7	1.72	2.7	–	–	10.37	88.1

[a] Assays based on homogenates of tissues from 50 synchronous third instar larvae
[b] Not quantitatively collected

The ontogenic profiles of these enzymes have been analysed using the same techniques of enzyme assay, partitioned
to individual enzymes on the basis of densitometric scanning
of stained PAG disc gels. Since the amount of soluble protein varies with age, as has been demonstrated in other
species of *Drosophila* (Sheehan *et al.*, 1979; Church and
Robertson, 1966), the results are presented as activities
per individual, rather than specific activities. Figure 1
shows the developmental pattern for *Esterase-1* and
Esterase-J. Considering *Esterase-1* first, activity rises
steadily from the time of hatching, and peaks in mid-pupal
life, at about the time that features of adult morphology
first become clearly visible. Activity appears to rise *via*
a series of plateaux which correspond to the transition from
one developmental stage to the next, and early development of
the next stage. In adult males, activity is initially high,
but rapidly declines to a stable level. The pattern is
slightly different in adult females, where activity remains
high over the first ten days of life, a period corresponding
to the period of peak egg production. Activity then declines
to the same resting level as the males.

The developmental pattern for *Esterase-J* is restricted to
a brief period from mid third larval instar to the time of
eclosion. It shows two clear peaks, one at the time of pupation, and another late in development of the adult within the
pupal case. There is no vestige of *Esterase-J* activity in
six hour adults.

Figure 2 shows the pattern of activity for *Esterase-2*.
This enzyme is present throughout the whole of development.
The pre-adult stages show three distinct peaks, approximately
corresponding to the transition from second to third instar,
third instar to pupa and late development of the adult.
Post-eclosion, there is a broad peak of activity over the
first ten days of adult life, after which the activity declines to a lower level, though it is not clear whether this
represents a stable plateau.

III   General Comparative Biochemistry

Since we do not know at this stage the natural substrate
or substrates of these enzymes, their activity has been
examined on a diverse group of esters, with the assumption
that compounds which are hydrolysed more efficiently may have
some structural similarity to the natural substrate. This
need not necessarily apply, but it serves as a guide for
future research.

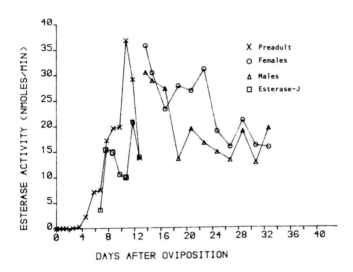

FIGURE 1.   *Developmental patterns of Esterase-1 and Esterase-J expressed as activity per individual.*

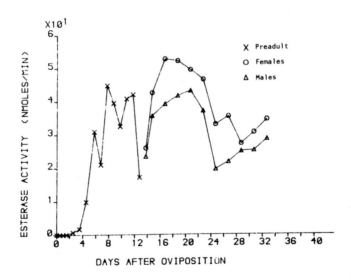

FIGURE 2.   *Developmental pattern of Esterase-2 expressed as activity per individual.*

A. *Substrate Specificity*

a. *Effects of changing the alcohol moiety*. The effects on enzyme activity of changing the alcohol group of the ester have been examined for a number of compounds of diverse structure (Table II). The compounds chosen are not particularly comprehensive, but cover a variety of sizes and shapes of molecules. Two observations may be made, firstly *Esterase-1*, though showing a very broad substrate specificity does not hydrolyse any of these compounds at a particularly high rate, and indeed with the exception of 2-naphthylacetate most compounds are hydrolysed at similar rates. Preliminary attempts to do kinetic studies using 2-naphthylacetate as substrate consistently resulted in curvilinear double reciprocal Michaelis-Menten plots. This is unusual in that many workers have assumed Michaelis-Menten kinetics to determine K_m values for non-specific esterases (van Asperen, 1962; Townson, 1972; Kapin and Ahmad, 1980), and we have experienced no problem with *Esterase-2* in this regard. A possible explanation of the failure of this enzyme to obey Michaelis-Menten kinetics may be simply that the substrate used is a very poor model for the true substrate.

TABLE II.   *Effects on Enzyme Activity[a] of Changing Alcohol Group*

	Specific activity[a]	
	Esterase-1[b]	*Esterase-2*[a]
Ethylacetate	46.6	N.D.[b]
2-Methyl-butyl acetate	39.1	38.2
o-Nitrophenyl acetate	4.9	45.3
p-Nitrophenyl acetate	67.8	166.2
2-Naphthylacetate	130.5	359.6
1-Naphthylacetate	69.5	785.8
Cholesterol acetate	N.D.	N.D.
Cholesterol oleate	N.D.	N.D.
L-Phenylalanine ethyl ester	10.22	13.37
L-Tyrosine ethyl ester	15.06	23.67

[a] *Specific activity : nmoles/min/mg protein*
[b] *N.D. = not detectable*

The second observation is that *Esterase-2*, which also has a very broad substrate specificity, does appear to show a discernable trend, at least on the series of acetate esters. Activity increases with increasing molecular weight of the alcohol moiety (subject to some steric restraints), and this increase may be associated with a decrease in K_m, since the K_m (app.) for p-nitrophenylacetate is about $40 \pm 6\mu M$, and for 1-naphthylacetate is about $25 \pm 2\mu M$. However, this increase in activity does not extend to the very large, hydrophobic cholesterol ester. It is possible then that the natural substrate (or substrates) for this enzyme have some sort of cyclic, or at least rather bulky alcohol group.

    b. Effect of changing the acid moiety. This study utilized a series of esters of p-nitrophenol, in which the number of acyl carbon atoms ranged from C=2 (acetate) to C=8 (caprate). *Esterase-1* shows a clear peak at C=3 (propionate), and *Esterase-2* at C=4 (butyrate). Comparison of four allozymes produced by the *Esterase-2* locus (Figure 3) reveals very little difference between allozymes. A possible exception is the enzyme produced by *Esterase-2^c*, where there appears to be no particular difference between the butyrate, valerate and caproate esters.

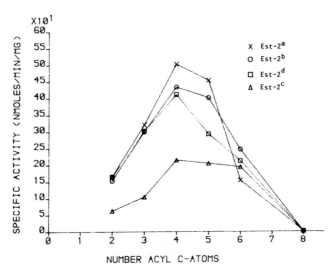

FIGURE 3. Effect of acyl group carbon chain length on Esterase-2 activity.

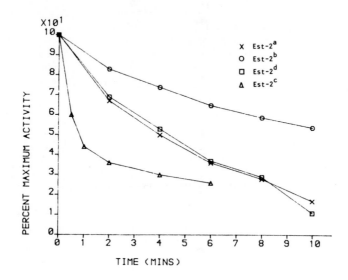

FIGURE 4.   *Thermal stability of Esterase-2 allozymes at 50°C.*

B. *Thermostability*

Temperature dependent differences in activity between allozymes have been implicated as a factor in the maintenance of an esterase polymorphism in the fish *Catostomus clarkii* (Koehn, 1969), and Narise (1973) reported differences between esterase allozymes in thermal stability at 60°C. A comparison of the thermostability of *Esterase-1* and *Esterase-2* of *D. buzzatii* at 50°C showed that *Esterase-1* is clearly a more stable enzyme, at least under the *in vitro* conditions employed. A more interesting comparison, however, is that of the thermostability of four allozymes at the *Esterase-2* locus (Figure 4). There appear to be quite dramatic differences, *Esterase-2*[b] being much more stable, and *Esterase-2*[c] much less stable than the other two forms.

C. *Kinetic Study of* Esterase-2 *Allozymes*

Somero (1978) has argued strongly for a more realistic approach to the study of biochemical adaptation. The parameters by which we characterise enzymes, viz. K_m, V_{max} and k_{cat} are not constants, but vary in response to environmental changes, such as temperature and pH fluctuations. This may be quite important in ectothermic organisms, where body temperature varies with environmental temperature, and pH is believed to vary in a compensatory manner (Malan *et al.*, 1976; Heisler *et al.*, 1976) so as to maintain a constant relative alkalinity. The study reported below is modelled closely on the elegant work of Place and Powers (1979) in their study of the *Lactate dehydrogenase* B allozymes of the fish *Fundulus heteroclitus*.

A preliminary analysis of *Esterase-2* allozymes is presented in terms of the pseudo first-order rate constant V_{max}/K_m, which is related to the specificity constant k_{cat}/K_m, the biochemical significance of which has been extensively documented (Brot and Bender, 1969; Bender and Kezdy, 1965; Cleland, 1975). A number of theoretical studies (Fersht, 1974; Crowley, 1975; Cornish Bowden, 1976) have suggested that k_{cat}/K_m should be large if an enzyme is to respond rapidly to changes in the level of substrate.

The kinetic data have been analyzed using the parametric, weighted non-linear regression method of Cleland (1979). The parameters K_m and V_{max} and the pseudo first-order rate constant V_{max}/K_m were estimated at each combination of four values of pH: 6.50, 7.00, 7.50 and 8.00 and three temperatures: 10°C, 25°C and 40°C, using 1-naphthylacetate as substrate (Table III).

TABLE III. Effects of pH and Temperature on V_{max}/K_m for Esterase-2[a]

Enzyme	Temperature (°C)	pH			
		6.50	7.00	7.50	8.00
Esterase-2[a]	10	7.66	12.02	9.89	8.71
	25	20.14	25.71	28.13	6.55
	40	26.65	34.88	43.27	32.65
Esterase-2[b]	10	5.53	8.69	6.18	10.39
	25	21.40	27.98	24.47	4.49
	40	26.29	37.24	31.10	37.64
Esterase-2[c]	10	1.98	4.50	3.20	2.41
	25	8.50	13.60	12.85	4.90
	40	8.49	11.09	16.20	8.27
Esterase-2[d]	10	4.31	8.97	5.02	6.59
	25	13.85	18.08	22.91	3.31
	40	15.90	24.35	52.20	27.70

[a]*All assays contained enzyme blanks to control for spontaneous hydrolysis of substrate.*

TABLE IV.   *Anova for Effects of pH and Temperature on*
            V_{max}/K_m for Esterase-2

Source of variation	d.f.	M.S.	F
Genotype (G)	3	436.22	21.93^b
pH (P)	3	208.34	10.47^b
Temperature (T)	2	1693.15	84.61^b
G x P	9	24.15	1.21
G x T	6	80.05	4.02^a
P x T	6	117.48	5.91^a
G x P x T	18	19.89	

$^a p < 0.01;$ $^b p < 0.001$

Comparisons between allozymes of the absolute values of V_{max}/K_m are not particularly meaningful because of the relationship, $V_{max} = K_{cat} \cdot E_0$. Thus we cannot say whether the observed differences in V_{max} between allozymes are due to differences in catalytic efficiency (k_{cat}) or in amount of enzyme (E_0).

In very general terms V_{max} and K_m both increase with increasing temperature, and V_{max} decreases and K_m increases as pH moves away from the optimum (7.0–7.5). However, an inspection of the data in Table III shows that the different allozymes do not always react in the same way to environmental changes. To try to determine the nature of these differences, the data were subjected to analysis of variance (Table IV). As expected, there are highly significant genotype, pH and temperature effects. What is more interesting is the significant genotype x temperature interaction. If the catalytic efficiency of different genotypes changes in different ways in response to alterations in temperature, then the potential for natural selection, mediated through fluctuations in temperature, to act on the *Esterase-2* locus must at least be reckoned as a possibility. This notion is attractive in that temperature has been implicated as a factor relating to polymorphism at the *Esterase-2* locus (*c.f.* Introduction).

IV   Possible Functions of *D. buzzatii* Esterases

It may be premature to speculate too much on the possible function of *Esterase-1* and *Esterase-2* in *D. buzzatii*. However, it has some heuristic value in the design of critical

experiments, and so the sections which follow represent an
attempt to weld the observations above into models which may
form the basis of testable hypotheses.

A. Esterase-2 *Function*

A number of possible functions have been suggested for
esterases, but given that *Esterase-2* is located primarily in
the alimentary tract some role in digestion or detoxification
seems plausible.

When PAG disc gels are stained for lipase-like activity,
there is an enzyme which occurs in the 100,000xg supernatant,
(with an R_f = 0.13, very different from the non-specific
esterases), which is active in the hydrolysis of triacyl- and
to a lesser extent monoacyl-esters of polyoxyethylene sorbi-
tan.   This observation is in agreement with that of Turunen
and Chippendale (1977) working on *Diatraea grandiosella*,
where they found one predominant lipase-like enzyme, with an
R_f quite different from the esterases.   It seems unlikely
that *Esterase-2* is a lipase.

The main options remaining are the digestion and/or ab-
sorption of other esters (e.g., sterol or vitamin esters), or
some role in detoxification (e.g., volatile esters, other
natural xenobiotics such as secondary plant compounds, or
synthetic xenobiotics such as organophosphorus esters).

We have made a small preliminary study of the volatile
fraction of *Opuntia stricta* rots and some cactophilic yeasts
grown in monoculture on sterilised, homogenised cactus.   The
volatiles were extracted by headspace collection at 40°C,
from 25 grams of each of four natural rots and four yeast
cultures, and were examined by gas chromatography and mass
spectrometry.   The study was only qualitative, and the 23
major volatiles from the eight samples were identified.   The
results are summarised in Table V.   There were three classes
of volatiles; alcohols, esters and ketones.   The four natural
rots had qualitatively similar GC profiles, but with some
differences in relative quantities.   The four yeast cultures
also were similar.   However, the GC profiles for natural rot
vs yeast culture were radically different (Table V).   Yeasts
produced alcohols and the esters derived from them, but
natural rots contained alcohols and primarily the ketones
derived from them, with very low levels of esters.   This
observation, combined with the relatively low level of
activity of the enzyme on low molecular weight volatile
esters suggests that activity on these compounds is not the
physiological function of this enzyme.

TABLE V.   *Volatile Components of Opuntia Rots*

Class	Compound	Natural rot	Yeast culture
Alcohol	Ethanol	+ [a]	+
	Propan-1-ol	+	+
	Propan-2-ol	+	+
	Butan-1-ol	+	−
	2-Methylpropan-1-ol	+	+
	Butan-2-ol	+	−
	2-Methylbutan-1-ol	+	+
	Pentan-2-ol	+	+
	Pent-3-enol	+	+
Ester	Ethyl acetate	+	+
	Ethyl propanoate	−	+
	Propyl acetate	−	+
	2-Methylpropyl acetate	−	+
	2-Methylbutyl acetate	−	+
	Furfuryl acetate	−	+
	β-Phenylethyl acetate	+	+
Ketone	Acetone	+	+
	Butan-2-one	+	−
	Pentan-2-one	+	−
	Diacetyl	+	−
	3-Hydroxybutan-2-one	+	−
	Nonan-2-one	+	−
	Acetophenone	+	−

[a] (+) indicates presence, (−) indicates absence of detectable quantities.

Perhaps the most attractive option is that this enzyme has some role in the detoxification of non-volatile xenobiotics. The substrate specificities shown in Table II reveal much higher activities on compounds with a cyclic group. Many of the known, naturally occurring insecticides and insect antifeedants have this type of structure (see for example Nokanishi, 1980; Bowers, 1980), and indeed many are esters.

The generally high levels of V_{max}/K_m shown in Table III indicate that this enzyme has the capacity to respond quickly to changing levels of substrate, which would be a desirable feature of a detoxifying enzyme, and the low K_m (app.) values exhibited for 1-naphthylacetate suggest that it would be efficient at binding low levels of toxic compounds.   There are currently experiments underway in this laboratory, attempting to use naphthyl esters as model toxins to test this hypothesis.

If we are to pursue this theory further, we will require information on potential toxins in the feeding substrate. Our knowledge of the chemistry of *Opuntia stricta* is negligible, but the work of Rockwood-Sluss *et al.* (1973) may provide a lead for future research. They reported correlations of allele frequency variation at the *Esterase-c* locus of *D. pachea* with changes in the level of alkaloids and non-saponifiable lipids. *Esterase-c* is the major 1-naphthylacetate hydrolising enzyme in this species. Alkaloids are well recognized for their toxic and pharmacological effects, and some have been shown to be hydrolysed by esterases in other species (Robinson, 1979). In this context, the identification of β-phenylethylacetate (Table V) in *Opuntia* rots is of interest. Although this compound may be simply a by-product of aromatic amino acid metabolism, it also has a close structural relationship to the β-phenylethylamines. These compounds form a major class of the alkaloids extracted from other species of *Opuntia* (Meyer, 1979).

B. Esterase-1 *Function*

Largely because of its developmental pattern, and its unique location in the haemolymph, I would like to propose that *Esterase-1* may act in conjunction with *Esterase-J* to modulate the titre of juvenile hormone. When one stains for esterase, despite being able to resolve 15-20 zones of activity, only *Esterase-1* and *Esterase-J* preferentially hydrolyse the 2-naphthyl ester. A possible explanation of this may be that their natural substrates are similar. The hormonal control of insect development has been reviewed by Riddiford and Truman (1978). It is the secretion of the steroid ecdysone which determines that a moult will take place. However, it is the concentration of juvenile hormone (JH) which determines whether the moult will be larval-larval, or larval-pupal. In holometabolous insects, although the transition from pupa-adult requires no JH, nonetheless JH is required to prevent premature development of adult organs in the pupa. Juvenile hormone is also required to control the maturation and maintenance of the adult reproductive system and adult tissue.

The JH titre declines throughout development and the level of JH is determined conjointly by the level of production by the corpora allata and by the rate of degradation by

JH esterases. That JH esterase is an important step in this
process is indicated by the fact that the inactive JH-acid is
by far the most common JH metabolite found in insect blood.

A consideration of Figure 1 reveals that *Esterase-J* has
precisely the developmental profile expected of a JH specific
esterase. It first appears in mid-late final instar larvae,
peaks at the larval-pupal transition, then peaks again late
in pharate adult development. What separates it from most JH
specific esterases is its localisation in the cuticle rather
than the haemolymph. This does not necessarily pose a problem, since at least one of the primary target cells for JH is
known to be the cuticular epidermis, and if it is necessary
to rapidly eliminate JH at the time of pupation, a JH esterase present in that tissue would provide an efficient mechanism.

However, for the efficient removal of JH from the entire
organism it would clearly be desirable to have a haemolymph
enzyme to remove circulating JH. *Esterase-1* may be a suitable candidate for such a role, since Figure 1 shows that its
titre is rising over a period when JH titre is falling, and
it peaks dramatically at a time when it is presumably desirable to rapidly eliminate JH. Furthermore, it is present in
the adult, where JH is known to have a function in other
insects, and therefore could help to regulate JH titre.

This hypothesis has the advantage of being amenable to
critical examination.

V   Future Research

The principal objective for the future will be the design
of *in vitro* and *in vivo* experiments to test the hypotheses
outlined above. If progress can be made towards the definition of physiological function, then knowing the feeding and
breeding sites will enable us to search for those factors
which might be responsible for the maintenance of polymorphism at these loci. The *Esterase-2* enzyme is particularly
interesting in this regard since we have demonstrated genotype x temperature interaction. Using similar techniques it
should be possible to test for genotype x substrate interactions. If these were significant, then we may be making
some progress toward an explanation of the high level of
heterozygosity at this locus. In Australian populations,
four alleles commonly have frequencies greater than 0.1.
Somero (1978) has stated: "an apparently important factor in
determining whether a single protein form or a set of isozyme

forms is required is the variability in substrate concentration faced by the particular type of protein under all environmental and life-stage conditions". If an enzyme faced a large variability in substrate types, as a detoxifying enzyme may well do, then there is no reason why a similar argument might not apply to allozymes.

ACKNOWLEDGMENTS

I would like to thank Drs J.S.F. Barker and D.C. Vacek and Mr. Minh Le for discussions and suggestions throughout this study. The volatile ester analyses were made possible by the assistance of Dr. F.B. Whitfield at C.S.I.R.O. Division of Food Research, North Ryde, N.S.W., Australia.

REFERENCES

Asperen, K. van. (1962). *J. Insect Physiol. 8,* 401.
Barker, J.S.F. (1981). *In* "Genetic Studies of *Drosophila* Populations" (J.B. Gibson and J.G. Oakeshott, eds.), p. 161. Australian National University Press, Canberra.
Barker, J.S.F. (1982). Chapter 14, this Volume.
Barker, J.S.F., and East, P.D. (1980). *Nature, Lond. 284,* 166.
Barker, J.S.F., Toll, G.L., East, P.D., and Widders, P.R. (1981). *Aust. J. Biol. Sci. 34,* 613.
Bender, M.L., and Kezdy, F.J. (1965). *A. Rev. Biochem. 34,* 49.
Bowers, W.S. (1980). *In* Insect Biology in the Future " "VBW 80" ' (M. Locke and D.S. Smith, eds.), p. 613. Academic Press, New York and London.
Brot, F.E., and Bender, M.L. (1969). *J. Am. chem. Soc. 91,* 7187.
Church, R.B., and Robertson, F.W. (1966). *J. exp. Zool. 162,* 337.
Cleland, W.W. (1975). *Acc. Chem. Res. 8,* 145.
Cleland, W.W. (1979). *In* "Methods in Enzymology", Vol. 63, p. 103. Academic Press, New York and London.
Cornish-Bowden, A. (1976). *J. molec. Biol. 101,* 1.
Crowley, P.H. (1975). *J. theor. Biol. 50,* 461.
Fersht, A.R. (1974). *Proc. R. Soc., Ser. B 187,* 397.
Heisler, N., Weitz, Ḧ., and Weitz, A.M. (1976). *Respir. Physiol. 26,* 249.

Johnson, F.M., Kanapi, C.G., Richardson, R.H., Wheeler, M.R., and Stone, W.S. (1966). *Univ. Texas Publ. 6615*, 517.

Kapin, M.A., and Ahmad, S. (1980). *Insect Biochem. 10*, 331.

Koehn, R.K. (1969). *Science 163*, 943.

Malan, A., Wilson, T.L., and Reeves, R.B. (1976). *Respir. Physiol. 28*, 29.

Meyer, B.N. (1979). M.Sc. Thesis, Purdue University.

Mulley, J.C., James, J.W., and Barker, J.S.F. (1979). *Biochem. Genet. 17*, 105.

Narise, S. (1973). *Jap. J. Genet. 48*, 119.

Nevo, E. (1978). *Theor. Populat. Biol. 13*, 121.

Nokanishi, K. (1980). *In* "Insect Biology in the Future " "VBW 80" " (M. Locke and D.S. Smith, eds.), p. 603. Academic Press, New York and London.

Place, A.R., and Powers, D.A. (1979). *Proc. natn. Acad. Sci. U.S.A. 76*, 2354.

Riddiford, L.M., and Truman, J.W. (1978). *In* "Biochemistry of Insects" (M. Rockstein, ed.), p. 307. Academic Press, New York and London.

Robinson, T. (1979). *In* "Herbivores: Their Interaction with Secondary Plant Metabolites", Ch. 11 (G.A. Rosenthal and D.H. Janzen, eds.), Academic Press, New York and London.

Rockwood-Sluss, E.S., Johnston, J.S., and Heed, W.B. (1973). *Genetics 73*, 135.

Sasaki, M., and Narise, S. (1978). *Drosophila Inf. Serv. 53*, 123.

Sheehan, K., Richmond, R.C., and Cochrane, B.J. (1979). *Insect Biochem. 9*, 443.

Somero, G.N. (1978). *Annu. Rev. Ecol. & Syst. 9*, 1.

Townson, H. (1972). *Ann. trop. Med. Parasit. 66*, 255.

Turunen, S., and Chippendale, G.M. (1977). *Insect Biochem. 7*, 67.

Watt, A.W. (1981). *In* Genetic Studies of *Drosophila* Populations", (J.B. Gibson and J.G. Oakeshott, eds.), p. 139. Australian National University Press, Canberra.

PART VI
SUMMARY
OF CONFERENCE WORKSHOPS

Summary of Conference Workshops

I Population Ecology and Dispersal

Panel members: A. Templeton, R. Mangan, S. Johnston, T. Markow, R. Richmond

A. Theoretical Considerations

Theoretical interest in the process and outcomes of dispersal has centered around the relationship between population structure and measures of inbreeding and variance effective population size. A major limitation of theory is its assumption that dispersal between populations results in effective gene flow. There is little hard empirical evidence on this point, and it emerged as one of the major deficiences of most studies of dispersal. Theory is often constrained by the lack of basic knowledge of population ecology, and many workshop participants urged that basic descriptions of dispersal parameters in natural populations are needed to provide a realistic baseline for theoretical studies.

B. Methodology

Fluorescent dyes (particulates) continue to be used widely for marking *Drosophila* and are available from two suppliers: Helicon Fluorescent Pigments and the U.S. Radium Corporation. The following colors have proven to be useful and can be used singly or in combination:

Yellow (2267), Green (3206), Blue (2205), Red (2225), Yellow/Orange (2220), and P-Green (2330).

Examination of marked flies is made most conveniently with a Mico-lite (Aristo Lamp Co.) attachment for dissecting microscopes. It is important to monitor the toxicity of these compounds for the particular species under study.

Neutron activation of whole flies was reported to be a new procedure which does not require field marking of organisms and may be useful to detect the origin of flies if they have fed on a test plant. Restriction enzyme analysis of mitochondrial and Y-linked sequences may prove to be very useful in tracing the dispersal of female and male associated genomes respectively.

341

ECOLOGICAL GENETICS AND EVOLUTION
ISBN 0 12 078820 9

C. Results

Studies of movement by *Drosophila buzzatii* between *Opuntia* plants has been successful in quantifying the amount and direction of dispersal. *D. buzzatii* showed substantial movement over 24 hr between *Opuntia* growths 100 m distant, and within 5-10 months colonized experimentally introduced *Opuntia* plants that were several kilometers removed from *Drosophila* populations, according to J. S. F. Barker. Further results suggest that increased population density enhances migration rates. Several discussants drew attention to empirical data which show that selection can blunt or greatly accentuate the genetic effects of dispersal.

D. Significance

Dispersal is not generally a random phenomenon. Genotype fitnesses may be maximized in habitats selected by their carriers. This possibility underscores the importance of studying the mechanisms and frequency of habitat selection. Dispersal propensity may differ between the sexes and respond to selection pressures associated with mating systems. Examples of dispersal which do not result in gene flow were presented, and the importance of measuring the genetic results of dispersal were stressed again.

E. Future Work

The need for a multidisciplinary, large scale study of dispersal in a single species was stressed. Discussion centered around the possibility of studying dispersal in *D. mojavensis* in an area characterized by isolated patches of host cacti and frequencies of chromosome inversions and allozyme alleles which differ from the main distribution area of the species. The necessity of applying the new procedures discussed above was emphasized. In summary, we agree that basic data on the extent and genetic significance of dispersal are critical if progress is to be made in attempts to understand basic evolutionary mechanisms in *Drosophila*.

II   Future Studies on Yeasts

Panel members: H. J. Phaff, J. S. F. Barker, W. T. Starmer, D. Gilbert, and D. L. Holzschu

A. *Background*

Recognition of cactus-specific yeasts dates from about 1974. Many of the species phenotypically resemble yeast species from other natural sources. Cactus-specific species were recognized by phenotypic properties and ecological associations. They were subsequently confirmed as species by determining nuclear DNA base composition and by DNA-DNA homology studies. Most yeast species that occur in cactus rots are cactus specific but not all. The majority has been described as new species and several more need to be described. Host plants and geographic separation play important roles in yeast evolution.

B. *Methods*

A description was given of the methods used to evaluate in a quantitative manner the yeast species that occur in cactus rots. Samples of rotting tissue are collected in sterile Whirlpack bags in the field. In the laboratory or other facility, one-gram samples of rot are placed in 9 ml of sterile water and serially diluted to 10^{-3}, 10^{-4}, and 10^{-5} of the original rot sample. Plating is done on yeast extract - malt extract agar acidified to pH 3.8 (with 1N HCl) to reduce bacterial growth. Counts of the different yeast colonies that develop on the plates in 3-5 days at room temperature are then made. A single colony of each type is picked and stored in a small screw-capped vial containing non-acidified malt agar. Selective media that contain specific carbon compounds used by only one or a few yeast species can also be used for quantitative enumeration of the different kinds of yeast occupying a rot. Mold growth is not a serious problem when plating cactus rot tissue. Yeasts consumed by adult *Drosophila* and larvae feeding in nature is done after sterilizing the exterior by immersing them in 70% ethanol for one minute and rinsing in sterile water. They can then be either homogenized and plated on agar medium or dissected and streaked with a loop on agar media as described above.

C. Dispersal and Ecology

The discussion indicated that there are many unanswered questions about dispersal of yeasts. Air currents are not considered important in yeast distribution. Vectors are largely unknown, although *Drosophila* probably play a role. Yeasts may be transmitted from rot to rot by adhering to the body parts of insects or by regurgitation from insect crops following an earlier feeding. In Australia *Cactoblastis* infestation most likely is responsible for microbial growth in *Opuntia* cladodes. Further work is also needed on transmission of yeasts through *Drosophila* eggs, larvae, and pupae. The possibility was raised that *Pichia cactophila* may be nutritionally adapted to dispersal by *Drosophila*, because it utilizes glucosamine, an amino-sugar possibly available during development of the fly. It was suggested that such co-adaptation requires detailed future study. A study of yeasts in South American cacti and *Drosophila* spp. was considered highly desirable, because some Australian Opuntias and their parasites came from S. America (probably Argentina). This would allow a comparison of the yeast floras on the two continents and a study of the effect of allopatry on yeast evolution. Another question requiring study are factors that restrict *D. aldrichi* within its host plant range. Experimental systems using cages and rotting cactus tissue with suitable controls (no bacteria and/or yeasts) were considered useful to study some of these problems.

D. Drosophila Nutrition

The question was brought up, "why don't adult flies live very long on rotting cactus in the laboratory (approximately 4 days)?" It was suggested that lack of sugar is a possible reason, but no proof was supplied. It may be that field rots are qualitatively different from laboratory rots or that adults have alternate sources of nutrients in the field. *D. buzzatii* appears to be an exception and is able to live longer than other species on rotting tissue. Bacterially infected tissues may play a role. We need to know more about the role of bacteria in cactus rots and their role of providing nutrients to other microbes (e.g., yeasts). An important lack of knowledge is how microorganisms decompose cactus tissue (both bacteria and yeasts) and how metabolic products of microorganisms can be used by other microorganisms and by *Drosophila* larvae. More information is needed on specific requirements that may be provided by microorganisms and toxins that may inhibit fly development. We need to know the role of bacteria and yeast independently of each other. Standardization of microbe strains for research is essential. Other areas that need further research include the influence of yeast strains of *Drosophila* genotypes, and determination of what individual yeasts in mono- or biculture do to fitness characteristics of the *Drosophila*. In Australia work will be continued on yeast-*Drosophila* relationships, effect of yeast on allozyme selection, and microheterogeneity in rots and its effect on larvae.

E. Future Studies on Yeast Systematics and Evolution

The following areas were considered desirable for study. We need to expand our knowledge of important cactus-specific yeasts by studies on DNA-DNA homology, DNA-ribosomal RNA homology, immunological comparison of selected enzymes, restriction patterns of nuclear and mitochondrial DNA. Since we can trace back cactophilic *Drosophila* outside the desert, the question can be asked if we can trace back cactophilic yeast communities to some tropical or temperate source outside the desert. The evolution of yeasts is intimately involved with the community in which they exist. How did the cactus yeast community originate? Are new species added or do species replace one another in response to the environment?

III *Drosophila* Systematics

Panel members: J. Ellison, A. Fontdevila, W. B. Heed, R. H. Richardson, F. de M. Sene, L. H. Throckmorton, M. Wasserman

Four questions emerged as central to the thinking of the group, although there was considerable overlap between areas, and some subjects could not be neatly pigeonholed.

A. *Further Study of the Repleta Group*

Why should further work be done on the systematics of the *repleta* group? In view of the already extensive systematic studies on the *repleta* group, it was thought pertinent to enquire whether much further work was really needed. The answer is unequivocal. A great deal of work, even of the most basic kind, remains to be done. There are species yet to be discovered and described. Major habitats are being destroyed by human activities before they have even been collected for species of the *repleta* group, and, since collecting methods have changed and improved greatly over the years, many geographic areas once thought to have been well-studied must be reinvestigated. In some cases, anatomical studies have been pursued only far enough to permit species description. Much valuable information remains to be collected on internal anatomy and reproductive structures of adults, and on features of the immature stages. Cytological analyses, both of salivary gland chromosomes and larval ganglion cells, are needed from many more samples of the group, adding information from new species as soon as possible and filling in details from species for which only fragmentary information exists. These will provide the minimum data needed for a more clear understanding of the evolution of the group. And this will be much improved as soon as molecular techniques are added to the approaches already in common use. Electrophoretic techniques have provided important information, and they should be extended, if possible, to include all species of the group. Studies of mitochondrial DNA are among the most promising new approaches, and they should be given highest priority for effort in the immediate future. Chromosomal DNAs should also be investigated as quickly as possible. At least selected single loci, or other specific chromosome regions, could be followed throughout the group, and this would add immensely, both to what we know of the group, and to general understanding of molecular evolution. Molecular techniques, especially gel electrophoresis, should be developed and exploited for species recognition and identification. The taxonomy of related species groups, both from Asia and from the Americas, should be extended to clarify present perspectives on the first appearance of the *repleta* group in the New World and its place of origin there. Finally, it is important that data be assembled on distribution of species, including altitudinal ranges, seasonality, and habitat selection. All of this together will provide a body of information that can give the most powerful support for future studies of the origin, evolution, and present dynamics of the *Drosophila*-yeast-cactus system.

B. *Importance of the Repleta Group*

Why should so much emphasis be placed on the *repleta* group? There are several groups of *Drosophila* that are large and diverse, and which undoubtedly have evolved in concert with communities of microorganisms. The *melanogaster* and *immigrans* groups of the Old World, and the *tripunctata* group of the New World, are among conspicuous examples that might be considered. None of these, however, have the advantage of being so well known cytologically. This, above all else, singles out the *repleta* group, for it provides a phylogenetic framework an order of magnitude better than is available for most evolutionary studies. Almost as important is the already extensive body of knowledge on the yeasts exploited by the *Drosophila* and on their interactions with host cactus species from groups whose evolutionary relationships are also well known. Biologists are well aware of the importance of the Hawaiian *Drosophila* as examples of explosive evolution on islands. The *repleta* group provides the major continental example for comparison and contrast to the Hawaiian case. It too has undergone explosive radiation exploiting the novel cactus-yeast community. But it has done this on the continents and so in many respects can be regarded as more typical of the general evolutionary case. Additionally, it provides examples of ongoing emigration and colonization that can be studied as they occur, and in this way it may provide sorts of study, and tractable materials, of a kind not accessible among the Hawaiian species.

C. *Immediate Needs in Repleta Systematics*

The following questions, while distinct in principle, tended to blur together, so their discussion is treated as a single unit here. What are the immediate needs of the systematists working on the group? What do other workers in the field need from the systematists? And what services should be provided to them by systematists? The immediate needs of the different systematists vary rather widely, depending on the particular regions where work is being done.

In most cases, priorities must be decided by the workers in
their own areas. Priorities affecting several areas have
already been pointed out in the answer to the first question
(Section A). One further item, critical to both systematists
and non-systematists alike, is that of the need for a catalogue of the *repleta* group. An undertaking of such importance requires more time and planning than could be devoted
to it in sessions of this type, but the panel felt that the
possibility of such a catalogue should be raised for future
discussion. This catalogue might contain descriptions of all
species, maps of distributions, general descriptions of the
ecology and habitats of each species, information on salivary
gland chromosomes and karyotypes, diagrams of male genitalia
and internal anatomy of both males and females, descriptions
of immature stages to assist nonspecialists in recognizing
these in the field, diagrams of relationships to assist molecular biologists (and others) in choosing species of most
use to them, and so on. It could, or should, also include
regional and locality keys to the species, illustrated to
make them as useful as possible to nonspecialists. In short,
it could bring together under one cover all pertinent information currently available on the group and make it readily
accessible to whomever might need it. It would also help
pinpoint gaps in our knowledge and fruitful areas for priority work in the future.

IV   Cactus Chemistry-Yeast-*Drosophila* Interactions

Panel members: J. Fogleman, D. Vacek, H. Kircher, and
A. Gibson

A. Overview

This workshop was intended to encompass all of the levels
of organization in the cactus - microorganism - *Drosophila*
model system including the interactions between levels. In
review, the system has a hierarchal structure and may be
generalized as: cactus → bacteria → yeast → yeast communities
→ *Drosophila* → *Drosophila* populations. Cactophilic flies
which use rots as substrates for both larval and adult stages
of the life cycle are affected by all levels of the hierarchy
as well as the interactions between levels. The lower organ-
izational levels reflect directly on the *Drosophila* population since the cactus substrate represents a major and meas-
urable part of the environment which impinges upon the gene
pool of the population. The biological significance of the

interactions cannot be over-emphasized. An examination of one level without consideration of the interactions of other levels loses considerable scientific focus.

It was pointed out that there are several areas of interest in this system which are not currently subjects of active research. Among these are the phylogeny of *Opuntia* on both continents, the taxonomy of the bacterial residents of cactus necroses, and the natural products chemistry of Australian cacti. Since research in these areas would benefit all, qualified specialists should be encouraged to pursue this work.

B. *Cactus Taxonomy*

Several members of the panel communicated their future research plans as they relate to this workshop. Art Gibson will concentrate his efforts on constructing a more parsimonious phylogenetic diagram for the tribe Pachycereeae, with greater precision between groups of closely related species. To accomplish this a broad data base will be necessary including collected flowers, fruits, seeds, and stems; scanning electron microscopic studies of seed, spines, stem cuticle, and pollen; and information on the occurrence of silica bodies, funicular pigment cells, calcium oxalate crystals, triterpene glycosides, and alkaloids. Extension of these studies into southern Mexico and the West Indies will contribute to the understanding of the evolutionary history of this tribe.

C. *Chemistry of Necrosis*

Since cacti must decay before they become substrates for *Drosophila*, the chemistry of the decay process is a relevant aspect of this model system. Chemical analysis of the cacti before and after rotting will help characterize the dynamic chemical milieu to which the flies are exposed. Henry Kircher is currently working on this project for the Cereus cacti in the Sonoran Desert. He expects to continue this study on *Opuntia* and indicated that analyses for the presence of sugars before and after decay would be particularly interesting since they are so important in the biology of all rot residents. In addition, he will continue his investigation of the effect of certain chemical classes (e.g. alkaloids, triterpene glycosides, and lipids) on both the microorganisms

and the *Drosophila*. One area of current interest is how the
rot organisms physiologically "handle" the short chain fatty
acids found in agria and organ pipe.

D. Cactus Influence on Yeasts and Flies

It was suggested that phenolic compounds in cacti may be
partly responsible for rot odors detected by humans and may
also be involved in host plant selection by the cactophilic
flies. Some work has already been done on phenolics in
saguaro, and their effects on *Drosophila* has potential for
future research.

The possibility that larval preference for yeast species
varies depending on whether cactus tissue has been added to
the yeast medium, was pointed out as one hierarchal interaction which should be considered in future experimental
design. Yeasts are chemically affected by the substrate
upon which they are grown. Experiments which test the
response of *Drosophila* to yeasts (nutrition, behavior, etc.)
should incorporate cactus tissue in the yeast medium in order
to more closely approximate the natural situation. The
cactus species used should be determined by the *Drosophila*
species being tested. One percent dry cactus powder or 10%
fresh tissue homogenate were recommended as standards.

V   Biochemistry of *Drosophila* Enzymes

Panel members: P.D. East, P. Batterham, J. Oakeshott, D.
Sullivan, R. Richmond

A. Alcohol Dehydrogenases

Most discussion throughout the workshop centred on alcohol dehydrogenases, and the evidence for the duplication
of the *Adh* locus in some members of the *repleta* group.
Apart from its interest as a taxonomic tool, the duplication
affords an opportunity to investigate evolutionary divergence
between and within the two loci. A number of areas where
more information is needed were identified:

1. The present estimate of the recombination distance
between *Adh-1* and *Adh-2* in *D. buzzatii* needs to be tested
and comparable data obtained for *D. mojavensis*.

2. Further work is in progress to determine the distribution of the duplication across species in the *repleta* group, and this should provide important information on the phylogeny of the group. However, in discussion it was revealed that the tissue distribution of the different ADHs and their electrophoretic patterns in *D. mojavensis* may differ from that of *D. buzzatii*. It will therefore be important to compare the developmental profiles and tissue distribution in those species shown to possess the duplicated locus.

3. A single pattern of activity on a gel does not establish that the locus is present as a single copy. It is only when duplicated loci give rise to electrophoretically distinct proteins that a duplication is detected by the techniques so far used with the *repleta* group. It will be desirable to analyse species subgroups believed to be closely related to the *mulleri* subgroup, but which apparently lack the duplication, for evidence of an electrophoretically cryptic protein or an *Adh* pseudogene.

4. There was also discussion on the necessity of treating enzymes as components of integrated systems, and the concommitant need to devise *in vivo* experiments to test theories on the physiological functions and adaptive significance of allozymes. In this context some planned experiments, to trace the metabolic fate of C^{13} labelled alcohols by nuclear magnetic resonance spectrometry using live *D. mojavensis* of various genotypes, were discussed.

5. It was noted that specific activity levels of *Adh* in the cactophilic species studied were considerably lower than in *D. melanogaster* even though the latter ostensibly has no duplicated *Adh* locus. It was suggested that a comparative study of the specific activities of ADH in the *mulleri* subgroup, and possibly some related species of the *repleta* group would be worthwhile, especially as information becomes available on the types and quantities of alcohols present in the breeding substrates. As a caution, however, attention was drawn to the lack of evidence of a relationship between variation at the *Adh* locus and fitness in ecologically relevant environments in *D. melanogaster*.

6. A final point of interest came from discussions comparing the *Adh* polymorphism in *D. buzzatii* and *D. mojavensis*. In *D. buzzatii* the polymorphic locus is *Adh-1*, which is expressed more in larvae than in adults. By contrast in *D. mojavensis* the *Adh-2* locus is polymorphic, and this locus is expressed relatively more in adults than in larvae. This dichotomy certainly merits further investigation, since it implies that <u>if</u> the *Adh* polymorphisms are being selectively

maintained, either the mode of selection is different in the two species, or the critical events occur in the larval phase.

B. Esterases

Future studies of function will depend to some extent on the results of work currently in progress with *D. buzzatii*. There are apparent homologies between some esterases in species of the *mulleri* subgroup, and it would be desirable to extend this information.

Given the background and the interesting results from the work of R.C. Richmond's group on *Esterase-6* in *D. melanogaster*, an investigation of the male specific ejaculatory bulb esterases of the *repleta* group would be well worthwhile. A suggestion that the ejaculatory bulb esterases of *D. mojavensis* may represent a duplication needs to be investigated further, and extended to other species of the *repleta* group. It was noted that *D. buzzatii* apparently lacks male specific esterases. Given their potential role in influencing reproductive performance, interspecific comparisons of the esterases in flies of the *repleta* group will be valuable.

C. Genetic Mapping

A problem which many encounter during studies on the cactophilic *Drosophila* is an inability to do genetic studies. There are no genetic maps, very few visible mutant markers, and only a small amount of information on the chromosomal location of protein markers. As a result, the extraction of genes and chromosomes is a protracted and inefficient process, and much worthwhile and important research is hindered.

D. A Cautionary Note

In the course of Workshop papers and ensuing discussion a recurring topic was the need to use appropriate experimental material for fitness tests in the laboratory. Inbred lines are of little use. It will be essential to investigate the physiological role of ADH before appropriate fitness tests of allozymes can be carried out. In this context, the lesson from studies with *D. melanogaster* was noted.

Index

2 3 4 5 6 7 8 9 0 1
A B C D E F G H I J